D1545163

# Archaic Cosmos

# Archaic Cosmos
## polarity, space
## and time

Emily Lyle

**Polygon**
EDINBURGH

© Emily Lyle 1990
First published in Great Britain in 1990
by Polygon
22 George Square,
Edinburgh EH8 9LF

Typeset in Linotron Sabon
by Koinonia Limited, Bury
Printed and bound in Great Britain
by Redwood Press, Melksham, Wiltshire
British Library Cataloguing
  in Publication Data
Lyle, Emily B (Emily Buchanan), 1932–
  Archaic Cosmos
  1. Western culture – sociological perspectives
  I. Title
  306.091821


ISBN 0 7486 6047 X

# Contents

# Contents

# Preface

I am grateful to the Bunting Institute at Harvard University, The Institute for Advanced Studies in the Humanities at the University of Edinburgh, and the Humanities Research Centre at the Australian National University for funded Fellowships in the period 1974-6 which was seminal for this work. I am also grateful to the University of Edinburgh for an Honorary Fellowship held in the welcoming and creative atmosphere of the School of Scottish Studies from 1970 until 1989 when the appointment became a Research Fellowship. Since the field of traditional cosmology is a new one which is only now beginning to win institutional support, I have often been dependent during the preparation of this book on the generosity of family and friends and would like to take this opportunity of offering warm thanks on my own behalf and on behalf of interested readers.

The chapters of this book first appeared as follows: chapter 1 in *Arv: Scandinavian Yearbook of Folklore* 36 (1980), pp. 127-31; chapter 2 in *History of Religions* 22 (1982), pp. 25-44; chapter 3 in *The Ballad Image: Essays Presented to Bertrand Harris Bronson*, edited by James Porter (Center for the Study of Comparative Folklore and Mythology, University of California, Los Angeles, 1983); chapter 4 in *Latomus* 43 (1984), pp. 827-41; chapter 5 in *Folklore* 95 (1984), pp. 221-3; chapter 6 in *Semiotica* 61 (1986), pp. 243-57, chapters 7, 8, 9, 10 and 13 in *Shadow* 1 (1984), pp. 22-8, 32-41, 3 (1986), pp. 3-8, 4 (1987), pp. 10-19, and 5 (1988), pp. 31-7; and chapters 11, 12 and 14 in *Cosmos* 1 (1985), pp. 1-14, 2 (1986), pp. 148-63, and 3 (1987), pp. 58-71. For permission to reprint I am most grateful to The Royal Gustavus Adolphus Academy, Uppsala; The University of Chicago Press; The Center for the Study of Comparative Folklore and Mythology in The University of California at Los Angeles; Editions Latomus, Brussels; The Folklore Society; Mouton de Gruyter; and The Traditional Cosmology Society.

My academic debts both to the living and the dead in the development of the wide-ranging thesis offered here have been many and varied; I hope that others in turn may like to draw upon this book in the future. I am very grateful for help in the preparation of the volume for the press to Peggy Morrison, to the illustrators, Leslie Davenport and Mairi Anna Birkeland, and to the Polygon production team.

Emily Lyle
School of Scottish Studies
University of Edinburgh

# Introduction

The chapters of this book require a word of explanation. They are adapted from articles published over the last decade and show the unfolding of concepts as my knowledge and understanding developed from early inklings in the light of my reading in a variety of areas. They contain a number of ideas that I believe are new in themselves or in their application, and, although it may be that in the not too distant future they will seem obvious enough, they have initially proved rather difficult for interested scholars to absorb. In my experience a fully developed methodology and the ease of communication it fosters result from informed debate within a subject area, and so I have concentrated on creating a field that I have called traditional cosmology in which this debate may take place. The methodology of this book is that of the individual enquiring mind asking questions, finding and articulating answers, checking them against new material, and going on to explore the new questions raised.

By the time the book had been completed, the enquiry had reached a point of rest on a number of issues. The apparent place of the human body in the cosmology became easier to understand when it was seen as the key metaphor in the system, allowing the positioning of both threefold and fourfold schemes in relation to it, so that asymmetries that initially seemed odd became quite natural and easily held in mind. The female single point that could be the whole, which was at first so puzzling although it seemed the only answer, became much easier to grasp when understood deictically as the point of anchorage in space and time. The trifunctionalism located by Dumézil in Indo-European class society was deeply enmeshed in a series of basic correspondences and this was more satisfactorily accounted for when it came to seem likely that it was rooted in pre-class society. I have not found, as Dumézil did, that the system was exclusively Indo-

European, and so have been free to incorporate into the enquiry other old world material which has helped very much in the elucidation of sovereignty. When the single principle of alternation – duality in action – was found to operate both for the kingship and for the calendar, again the proposed system became more easily grasped than at an earlier stage when the question of the dark days was treated as an isolated problem. The distinguishing of three axes of polarity (see chapter 7) provided a framework which was to prove valuable in the Chinese as well as the Indo-European context.

The book as a whole challenges deeply held assumptions about the roots of our culture, and acceptance of the idea that the system which I have sketched in these chapters once existed in actuality would bring about a paradigm shift. I see no alternative to exploring the possibility that such a paradigm shift is called for. For too long we have been turning a blind eye to a mass of cosmological material that finds no place in current thinking. It is time to adapt our thinking in order to take full cognisance of the cosmological aspect of our cultural history.

# 1. Cosmos and Indo-European Folktales

The theory concerning the structure of the Indo-European cosmos that I am presenting in this chapter[1] coordinates two separate lines of thought which originated in the study of primitive classification published by Emile Durkheim and Marcel Mauss in 1903 (Needham, ed., 1963). It is a mere adjustment to the work of earlier scholars but its importance lies in the fact that it provides a bridge between two influential earlier theories and allows us to draw on both to establish a more complete pattern than either provides separately.

Following Durkheim and Mauss, Georges Dumézil, who has been responsible for a major development in comparative Indo-European studies, held that the structure of society was the basis of classification and, in a series of studies, argued that there was a tripartite division of Indo-European society into groups representing the three 'functions' (the sacred, physical force, and fertility) which were: 1 priests, 2 warriors, and 3 cultivators (Dumézil 1958; Littleton 1982). There is some question as to whether the social groups are more fundamental elements of classification than other triads such as those of colours and of the seasons (Gonda 1976: 125-7, 170), but there is general agreement that triads were indeed of major importance among Indo-European peoples.

In addition to this categorisation by three, a categorisation by four also stemmed from the work of Durkheim and Mauss. Mircea Eliade, in particular, has demonstrated the importance for mankind of orientation in space, with the polarities of right and left, in front and behind related to the four quarters. It is clear that Eliade regards his studies of the four directions round a centre as applying to the Indo-European peoples as well as to other groups on which he has focussed more attention. At the same time, he accepts Dumézil's arguments in favour of the tripartition of Indo-European society (Eliade 1978-86: 1. 3, 42-3, 190-5). We therefore have a situation according to which a people classifies both by three and by four. I feel that we cannot understand the complete system without putting the three and four into relationship with each other and I suggest in this chapter how this can be done.

An earlier attempt to relate the three and the four was made by Alwyn Rees and Brinley Rees in their book *Celtic Heritage* published in 1961 (see especially 112-33). They brought together the ideas of Dumézil and Eliade and, using an Irish example of four provinces round a centre, they related Dumézil's three social groups to three of the four directions. It is, of course, the fourth direction that is the problem and they solved it by adding a fourth social group, the serfs, which is not in Dumézil's triad, and placing that in the fourth quarter.[2] Although this is a sensible suggestion for which there is some support I do not think that it goes to the root of the problem and my view of the matter is that an understanding of the true relationship between the triad and the four quarters can be reached only through the introduction of the practice of adding the whole to the number of the components (cf. Gonda 1976: 8).

If the triad is completed by the whole, then we have three separate units and also a triple entity, the totality of the three units,[3] and Dumézil has made an interesting observation which enables us to identify this triple entity within his system. Although he was concerned primarily to distinguish the three functions, he became aware that, in addition to the gods of the three functions, there was a trivalent goddess who was related to all three. He defines her position in this way (1970: 300):

> In fact, among the Germans as well as the Indo-Iranians, we can observe the following structure. As a counterbalance to the group of masculine gods, each of whom embodies distinctly and analytically one and only one of the three basic functions, there is a goddess who synthesizes these functions, who assumes and reconciles all three, ...

If the goddess synthesises all three functions, it seems that she is attached to the population as a whole rather than to any particular social group. She is the deity of the entire people.

It should be observed that, since the goddess is present within the three functions, she need not be located separately and so her existence can easily be overlooked when triads alone are being considered. The special value of the relationship with the world quarters for students of the system lies in the fact that the requirement of relating to the tetrad of the directions means that the fourth element, the whole which completes the triad, must be made explicit. The goddess is brought out of hiding and becomes the deity of one of the four directions. In the case of the four provinces of Ireland studied by Alwyn and Brinley Rees, it seems that Leinster, which is linked with the goddess Brighid (Mac Cana 1983: 93), is the special province of the goddess, and I would replace the scheme in *Celtic Heritage* (which includes the additional social group of serfs) with the scheme in figure 1.1.

Ulster
warriors

Connacht                               Leinster
priests                          whole people (goddess)

Munster
cultivators

*Figure 1.1*

Associated with the three social groups are three colours: white for priests, red for warriors, and black for cultivators.[4] It seems that the goddess is associated either with all three of these colours, as in the description of an Indian goddess with white head, red upper body, and black lower body (Sastri 1974: 213), or alternatively with a colour not in the basic triad, generally yellow – the additional colour in the set of four for the Indian castes, or green – the colour for the goddess Flora at Rome (Dumézil 1954: 45, 54). The grouping of colours can be as shown in figure 1.2.

red

white                           trichrome, or yellow/green

black

*Figure 1.2*

Another fundamental triad is that of the divisions of the year. The

Indo-Europeans had three seasons – summer, winter and spring – but at the completion of a cycle of the seasons there was a brief period apart from the seasons which was held to reflect the year as a whole; i.e. although the year had three seasons it had four parts – the three seasons and the year.[5] In Ireland, this out-of-season period fell at Samain, at the beginning of the dark half of the year (Mac Cana 1983: 127-8; Rees and Rees 1961: 84, 88-91). The cycle of the three seasons with the year itself as encompassing fourth can be shown as in figure 1.3.

summer

spring                                                              year

winter

*Figure 1.3*

What bearing does all this have on folktales? That remains to be discovered, but if correspondences like the basic examples considered here were operative at an early period then we would expect at the very least to find some traces of these ideas embedded in traditional narratives. One motif which seems to lie particularly close to the concept of the three parts and the whole is that of the threefold death. Donald J. Ward has already discussed the three modes of death associated with the three functions – hanging, burning or being struck with a weapon, and drowning – as possible reflections of actual sacrificial practice and he has drawn attention to a death brought about in all three ways at once. He comments that it is probably significant that in Irish instances the threefold death takes place at a festival as this would be an appropriate time for sacrifice.[6] He does not refer to the fact, but the festival mentioned in his sources is Samain, the time according to the scheme suggested here when it would be appropriate to offer sacrifice to the triple goddess, representing the whole.

An Irish story I shall look at briefly in conclusion deals with a mother (Boand, the River Boyne) who gives birth to three sons,[7] a grouping which could clearly be an illustration of the idea of the encompassing female as the whole and the three males as the distinguished parts. The three sons are named from the mother's experiences as she gives birth. The first is Crying Music, from the pain she felt, the second is Laughing Music, from her joy at the birth of two sons, and the third is Sleeping Music, from her exhaustion following the birth. In a nineteenth-century study, Eugene O'Curry put the three types of music into relation with the three seasons (1873: 3.217), and, in the light of the present theory, we would expect this kind of correspondence. Two of the ideas are polar opposites – Crying Music / Laughing Music – while the third, the Sleeping Music which is elsewhere spoken of as a healing music, may be a mediating term between the two extremes. The equivalent seasons, then, would be the opposites – winter : summer – with spring as the mediating term. The order of birth is of special interest. The oldest son is connected with winter, the second with summer, and the youngest with spring. From the schemes above, we can see that the youngest is also associated with white (which could be symbolised by his having fair hair) and with the priesthood (which could be associated with his making of virtuous choices). We seem here to have accord between the structure of the cosmos and the motif of the three brothers, the youngest of whom succeeds in the quest.

# 2. Dumézil's Three Functions and Indo-European Cosmic Structure

The Indo-Europeans, as Eliade has recently reaffirmed in *A History of Religious Ideas* (1978-86: 1.190), were accustomed to cosmicising space, and yet, when we consider exactly how they did so, we find little current discussion. This may be because Georges Dumézil has explicitly dissociated his system of the three functions from any total cosmic scheme involving such basic dimensions as space and time. I think this self-imposed limitation is unfortunate and is now hindering our fuller understanding of Indo-European religious structures, and I will attempt here to redress the balance by looking at some of Dumézil's findings in the light of the concepts of vertical and horizontal space and, more briefly, of cyclic time.

To understand why Dumézil takes the position he does, it is necessary to go back to see how, after a period of fumbling, he reached his main insights when he adopted a sociological approach. This process has been outlined by C. Scott Littleton as follows:

> Gradually ... as he became more and more sensitive to the importance of social organization in the study of religious phenomena – a sensitivity that, as noted elsewhere, seems to have been due in no small measure to his association with Mauss and Granet – he began to see a correlation between Indic and Iranian stratification patterns and arrived at the then still tentative conclusion that Proto-Indo-Iranian social organization was composed of the now familiar 'three functions': priests, warriors, and food producers. He was supported in this conclusion by the independent but parallel conclusions reached by Benveniste. Only later did he come to recognize the extent to which the major gods of the *Rig Veda* were personifications, 'représentations collectives,' of this tripartite social structure. Once this had been determined, he began to probe deeper and to examine other ancient I-E-speaking societies to see if a

similar pattern could be found. By the mid-thirties he had found in the French school of sociology, as developed by Durkheim, Mauss, Granet, and others, the theoretical basis he had been seeking; and, as more evidence was obtained, especially from the Latin and Norse regions, the system rapidly began to crystallize.[1]

Although the system was developed in detail as the years went by, it continued to include only social groups and deities, and this fact is rather surprising when we remember that Dumézil has expended considerable energy on exploring other triads, each element of which he relates to one of his three functions of the sacred, physical force, and prosperity. Placed in correspondence with the social groups (priests, warriors, and food producers), we are given, among other items, cosmic levels (heaven, atmosphere, and earth), body parts (head, body above the waist, body below the waist), and colours (white, red, and black or blue). That is to say that we have macrocosmic and microcosmic equivalents to the social groups on the mesocosmic level, and we also have colours which form a fundamental part of cosmic systems, like those of the Chinese and the Navajo Indians, for example (cf. Needham, ed., 1954-84: 2.262-3; Reichard 1950). Dumézil, however, argues that these triads are not a fixed part of the system he is studying. All that he has allowed in a recent statement of his position is that certain triads (in contrast to others with flimsy or transitory connections) do form a 'more stable' relationship with his three functions.[2] If there is a complete cosmic scheme, however, the relationship of fundamental parts is not just relatively stable; these parts must provide the constant correspondences on which the articulation of the whole depends. In fact, it does seem that Dumézil's studies have brought out a number of these constant correspondences, and, once we admit that other sets are as valid as the social one, we are free to see the correspondences as forming a structure which is not determined by the social plane.

The scheme shown in figure 2.1, derived from Dumézil's writings,

| | Macrocosm | Mesocosm | Microcosm |
|---|---|---|---|
| 1. The sacred ...... White | Heaven | Priests | Head |
| 2. Physical force ... Red | Atmosphere | Warriors | Upper body |
| 3. Prosperity ...... Black | Earth | Food producers | Lower body |

*Figure 2.1*

can serve as a useful introduction to this discussion of cosmic structure. In these instances both the body and the cosmos are seen vertically. There are three levels, one above the other, and this fits completely with the tripartite division found in the three functions. Here we have vertical triads that Dumézil has studied thoroughly. The only change I would propose initially is that we view all the sets of items as forming a part of the cosmic scheme on an equal footing.

The division of the horizontal plane presents greater problems in relation to Dumézil's three functions, for this plane is divided not into three parts but into four – the four quarters radiating from the centre. F.B.J. Kuiper suggested in 1961 that the Indo-Europeans had a classificatory system which arranged the phenomena 'according to the four points of the compass' and that Dumézil's system might be replaced by one along these lines (Kuiper 1961: 39). In an interesting development in the same year, Alwyn and Brinley Rees made a move not to replace Dumézil's system by a system of the four directions but to incorporate it in such a system.[3] They did not press the point that the cosmic scheme they were studying was larger in scope than Dumézil's system, but the idea that his system is part of a greater whole is implicit in their work. They saw that each of the three functions could be placed in correspondence not only with one of the superimposed levels of the cosmos but with one of the world quarters as well.

Part of the support for this idea came from India where the four castes which developed out of the three Indo-European social groups by the addition of a servant group (the *śūdra* caste) are found located in the four directions along with their appropriate colours. The castes in hierarchical order are shown in figure 2.2.

| | |
|---|---|
| *Brāhman* (= I priests) | : White |
| *Kṣatriya* (= 2 warriors) | : Red |
| *Vaiśya* (= 3 food producers) | : Yellow |
| *Śūdra* | : Black |

*Figure 2.2*

They occur in the same sequence round the four world quarters.[4] Figure 2.3 should be read clockwise starting from *Brāhman*.

Kṣatriya
Red

Brāhman　　　　　　　　　　　Vaiśya
White　　　　　　　　　　　Yellow

Śūdra
Black

*Figure 2.3*

The white and red quarters present no problem. We have the straightforward correspondences – *brāhman* (= priests) : white; and *kṣatriya* (= warriors) : red – and so can replace the *brāhman* and *kṣatriya* castes with the equivalent Indo-European groups in the same quarters. This is what was done by the Reeses, and I agree with them here. However, I differ from them when they continue to follow the circle of the castes, replacing the *vaiśya* caste with the food producers (farmers) and the *śūdra* caste with an additional group of serfs, which has no place in the triad of the functions. As the four castes are a later development, I think we should follow the evidence of colour rather than that of caste where the two diverge, and Dumézil has established that it was not yellow but black that was the colour of the food producers.[5] The colours alone can be set out as in figure 2.4.

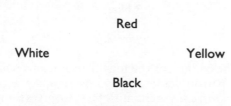

Red

White　　　　　　　　　　　Yellow

Black

*Figure 2.4*

If we then attach the social groups as studied by Dumézil to the appropriate colours, we arrive at the pattern of figure 2.5.

The quarter that has no equivalent among Dumézil's three functions clearly calls for special study. For an indication of what it contained, we can turn to the colours of the four chariot teams that raced at Rome, a subject that has been fully and sensitively studied

Red
2 warriors

White                                                          Yellow
I priests

Black
3 food producers

*Figure 2.5*

by Dumézil (1954: 52-6; 1969: 218-23). Cedrenus has the story that Romulus divided the city of Rome into four parts which he connects with the four elements to which he attaches the four colours: green, blue, red, and white.[6] We have here the basic triad of colours connected with the three functions – white, red, and blue (= black) – together with the supplementary colour, green, which is equivalent to the Indian supplementary colour, yellow. Placing the Roman colours round the world quarters as in the case of the Indian equivalents, we have:

2 red

I white                                                         Green

3 blue

*Figure 2.6*

These were the colours of the four chariot teams which also had their particular patron deities. In two overlapping accounts, John Lydus (*De Mensibus*, ed. Wuensch 1898: 4.30) gives the names of the deities linked to the colours as: white – Jupiter; red – Mars; green – Venus or Flora; and blue – Saturn or Neptune. These can be placed on the world quarters as in figure 2.7.

The three colours and quarters associated with the functions are linked with gods, while the fourth colour, with its corresponding quarter, is linked with a goddess. Lydus speaks of Flora as being Rome itself, and Dumézil adds supporting comments. It appears that the goddess is no particular function but the whole city. It is this

2 red
Mars

I white                              Green
Jupiter                        Venus or Flora

3 blue
Saturn or Neptune

*Figure 2.7*

conception of the whole – and the homeplace – that is located in the fourth quarter.

The Roman evidence is most important in establishing this point since it includes references to colours and the divisions of the city, but the tetrad made up of the three gods of the functions and a goddess of the homeplace occurs elsewhere. In the case of Athens, Dumézil (1958: 59) and Benveniste (1973: 235-7) have discussed a grouping of the three gods, Zeus, Poseidon, and Hephaistos, and the goddess Athene whose name is linked with that of the city. The statues of a comparable tetrad of deities accompany that of King Antiochus I of Commagene (69-34 BC) at his monumental tomb at Nimrud Degh. The monument is Greco-Iranian in inspiration, and the male deities have compound names derived from Greek and Iranian traditions: Zeus-Oromasdes, Apollo-Mithras-Helios-Hermes, and Artagnes-Heracles-Ares. The goddess, who is represented holding a cornucopia, is spoken of as 'my country Commagene which nourishes all'.[7] In these three cases, we have sets of three male deities and a female deity who is associated with the city or country – Rome, Athens, or Commagene.

Dumézil uses all these instances and is, of course, aware of the importance of the goddess. Interestingly, he sees her in two ways: (1) as trivalent deity, synthesising the three functions, and (2) as belonging especially to the third function in association with the twin gods, the Aśvins (Littleton 1982: 9, 15-6; Dumézil 1970: 300). The present study confirms that the goddess can be looked at in two ways, but the attention paid here to cosmic structure allows us to define her position more exactly.

The goddess as trivalent is active in all three functions, or, as one can say in terms of vertical space, she is active at all three of the superimposed levels of the cosmos, and so the vertical triad is not a simple one of three male components; it consists of the three male components and the female whole interfused in each part.

On the horizontal plane with its four directions, the goddess's presence is explicit and inescapable, and it is this that makes the study of the horizontal plane so valuable to those attempting to grasp the system. As we have seen, the goddess is one member of a tetrad and has a separate station of her own which follows the stations of the gods of the first and second functions and precedes that of the god of the third function. The tetrad with its four members can fall into halves, and, counting from the top of the hierarchy, one has, as the first half, the gods of the first and second functions, and, as the second half, the goddess and the god of the third function. This gives a clue to the nature of the goddess's special association with the third function. It is not, as Dumézil says, that she belongs to the third function herself but that she and the god of the third function share a half in opposition to the gods of the first and second functions. In the Roman instance, this can be shown as in figure 2.8. We may suspect that the lumping together of the goddess's attributes with those of the third function in Dumézil's scheme has led to a certain confusion and that some redefinition of the boundaries of the third function will be required.

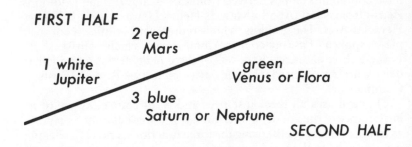

*Figure 2.8*

The fact that each of the three functions is represented by a single god in the tetrads that have just been discussed does not necessarily contradict Dumézil's findings about the representation of a function by a dual deity or a pair of gods (as Mitra-Varuṇa and the twin

Aśvins). As awareness of the structure grows more sophisticated, complexities are likely to emerge. What is apparent from the examples quoted, however, is that it is possible to have the three functions each represented by a single god, and it is this that seems appropriate at the level of the fundamental and elementary structure being discussed here, which is as in figure 2.9.

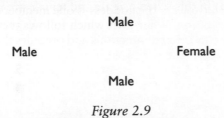

Figure 2.9

The idea of a male half and a female half is familiar, but this unequal division is less so.[8] Why should three of the world quarters belong to the male and only one to the female? When I was first considering the matter, it seemed to me that Jung's claim (1916: 227-8) that the male trinity was phallic in origin ought perhaps to be introduced to account for the male's triple share. I then came on a concept of the relationship between the world quarters and the genitalia as understood among the Hausa of West Africa,[9] which so perfectly expresses the position arrived at here that it seems likely to be at the root of the Indo-European structure whether or not the similarity of ideas is to be explained in terms of cultural connection. Among the Hausa, the horizontal plane, as seen in figure 2.10,

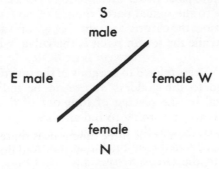

Figure 2.10

is divided into halves along sexual lines, east and south being regarded as male and west and north as female. At the ceremony of the
naming of a child, however, one quarter within the female half is
marked out as belonging particularly to the female. The child faces
east, and heaps of ash are placed on the ground in front of it, three
for a boy and four for a girl, to form a triangle or quadrilateral figure
as in figure 2.11. In this scheme, male and female are both found in

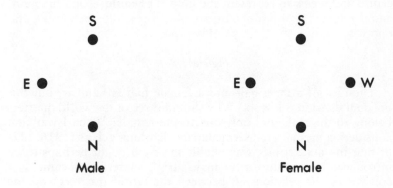

*Figure 2.11*

three of the four directions (south, east, and north), while the female
alone is found in the fourth (west). The scheme closely resembles the
Indo-European one where the goddess is related to all three functions
but has a separate place in her own quarter. The Hausa explanation
is by reference to the sexual parts: the male and female have three
points in common, the clitoris and labia being equivalent to the penis
and testicles, while the fourth point is the vulva, which is uniquely
female.

The three quarters related to the gods of the functions and the one
quarter related to the goddess appear to be indicated in Indo-
European ritual by the placing of four out of the five sacrificial
animals, which are, according to Indian sources, man, horse, bull,
ram, and he-goat (*Śatapatha Brāhmaṇa* 6.2.1.18). The last three of
these are grouped together in the Indian *sautrāmaṇī* sacrifice which
is equivalent to the Greek *trittyes* and the Roman *suovetaurilia*,
where the three animals sacrificed are bull, ram, and boar, the Indian

he-goat being considered a substitute for the boar. Dumézil and Benveniste both concluded that each of these three animals represented one of the three functions and that the sacrifice of the three together 'united symbolically the three orders of society'.[10] These scholars, in neglecting the evidence of the horizontal plane, did not discuss the Indian sacrifice of the raising of the fire altar, the *agnicayana*, in connection with the *sautrāmaṇī*. The same three animals reappear but this time in three of the directions and associated with the other two sacrificial animals, man and horse. The man is placed in the centre and seems to represent the king. The ram (second function) and the he-goat (third function) are placed opposite each other, and opposite the bull (first function) is placed the horse (figure 2.12).[11]

2 ram

I bull                                                    Horse

3 he-goat
(= boar)

*Figure 2.12*

It seems that placed in the quarter which has no equivalent function we have the theriomorphic form of the 'Indo-European mare goddess' recently studied by Wendy D. O'Flaherty (1979-80). The many-faceted goddess can, of course, be represented in various theriomorphic guises. She may be cow, ewe, or sow when matched with the gods of the three functions, but, as the above scheme indicates, in her own right she is mare, and we can see how appropriate it is that the marriage of king and goddess should have been represented theriomorphically as the mating of stallion and mare (Puhvel 1970: 159-72; 1987: 169-76).

Besides the sets of social groups, colours, deities, and sacrificial animals placed in the four directions, we can also examine a set of virtues which has been studied by Dumézil in relation to the concept of the world quarters. This set is of considerable interest in itself, but its special value in this context is that it has a connection with narrative material which will enable us to make a transition from the structure of the horizontal plane to a cosmogonic myth that offers an explanation of how it came to be that the male and female components of the cosmos were distributed as they were.

We can start with a look at the figure of the king. As Dumézil has pointed out, the king is held to have the virtues appropriate to all three functions: the ability to practise correct ritual for the first, courage for the second, and generosity for the third. We now have to think of the place of the goddess, and it seems to me that the king's sacred marriage may have been essential to his becoming king partly because it was through that marriage that he acquired the highest of his virtues – 'truth' – which I suggest was the special virtue of the goddess. The concept of truth or cosmic order has been much studied (see, e.g., Dillon 1975: 127-34), and I do not expect that what is said here will exhaust its meaning. It can readily be seen, however, that while each of the gods of the functions is concerned with one function only, the goddess, as relating to all the functions, has an overview and is in a position to strike a fair balance. On becoming king, a man must detach himself from a particular class and be fair and just towards all, seeing to it that each part has what it ought to have and ensuring too that each fulfils its function, does what it ought to do. This is how cosmic order is maintained within the kingdom on the level of society.

Plato's *Republic* is concerned with the nature of this order. It sets out to answer the question, What is justice? and eventually reaches the definition that it is each part doing what is appropriate to it, which is a clear statement of this kind of ordering (*Republic* 434-5; trans. Shorey 1930-5: 1.372-6). Much that precedes this definition in the *Republic* is also relevant to the present discussion. Plato's ideal city contains three hierarchically ordered classes – philosopher-rulers, soldiers, and farmers and craftsmen – which Dumézil has noted are related to the three functions (1958: 16; 1941: 257-61). In searching for justice, Plato affirms that his city, being perfectly good, will possess the four virtues of wisdom, courage, temperance, and justice. He takes these virtues in turn and matches wisdom with the philosopher-rulers, courage with the soldiers, and temperance with the farmers and craftsmen.[12] He then concludes that justice belongs to no particular class but to all three. As Plato not only has classes equivalent to the three functions but also treats the whole as a separate category with its distinctive virtue, it seems that he is working with a scheme which is either that of the four quarters as discussed here or analogous to it. It can be set out as in figure 2.13.

2 soldiers
Courage

I philosopher-rulers                                    All
Wisdom                                                Justice

2 farmers and craftsmen
Temperance

*Figure 2.13*

Dumézil's discussion of the world quarters and the virtues of the different social groups is included in his study (1973a) of a complex of stories where the Indian king, Yayāti, is a central figure. Since Dumézil had not defined the special nature of the fourth world quarter, he was not led to the conclusions that I draw here, but his explication of this group of stories has been invaluable, and I can refer the reader to his work for full discussion of the narratives. I concentrate exclusively on certain social, spatial, and genealogical elements within their structure.

By comparing two stories about Yayāti in the *Mahābhārata*, Dumézil brings out the point that this primal king can be seen to be distributing among his sons and grandsons both the different functions with their special virtues and the kingdoms lying in the four directions and at the centre. In the story of greater interest here – that of Gālava – the focus of attention is Yayāti's daughter, Mādhavī, and her four sons, each of whom has been begotten by a different father and has a special virtue. These virtues are summed up in a single sentence addressed to Mādhavī: 'You have borne one son who is royal in generosity, another who is a hero, a third who is dedicated to truth and Law, and one more who is a sacrificer.'[13] Dumézil has related these virtues to the three functions – generosity to the third function, heroic valour to the second, and both truthfulness and making of sacrifices to the first. Elsewhere in the same study, however, he recognises truthfulness as being a higher virtue than the others and as not belonging purely to a particular function (1973a: 29, 44-5), and J.A.B. van Buitenen distinguishes this virtue from the others in his discussion of the sentence quoted above. He accepts the relationship proposed by Dumézil between the three functions and three of the virtues – which can be expressed as: (1) religious zeal, (2)

heroic valour, and (3) generosity – but he does not include truthfulness among the virtues of the particular functions, suggesting instead
that 'perhaps this *satya-ṛta* is the value that overarches the three
classes' (*Mahābhārata* 3.172). That is to say that, without any
reference to Plato, van Buitenen suggests a scheme comparable to
that in the *Republic* (figure 2.14).

2 heroic valour

I religious zeal                                     Truthfulness (overarching
                                                           the three classes)

3 generosity

*Figure 2.14*

The narrative material in the *Mahābhārata* offers us for the first
time the opportunity to study the source of the components in this
fourfold scheme. In the story of Gālava, each son derives his special
virtue from his father, but the pattern being explored here suggests
that the virtue that overarches the others would be received through
the female, and I have incorporated this suggestion in the following
diagram. Above each of Yayāti's grandsons is shown his father, one
of the four lovers of Mādhavī, and above the grandson noted for his
truthfulness I have also placed Mādhavī herself (figure 2.15).

| Yayāti | | | |
|---|---|---|---|
| Viśvāmitra | Divodāsa | Mādhavī Auśīnara | Haryaśva |
| Aṣṭaka I religious zeal | Pratardana 2 heroic valour | Śibi truthfulness | Vasumanas generosity |

*Figure 2.15*

In the other Yayāti story discussed by Dumézil – that of the
transferred old age – Yayāti has five sons; the four disobedient ones
are banished to the periphery (which may be thought of as the four
quarters), and the obedient one, Pūru, succeeds Yayāti in his king-

dom (the centre) (*Mahābhārata* 1.190-4). This distribution gives four sons who correspond structurally to the four grandsons in the other story and also a fifth in the central position (figure 2.16).

| Yayāti | | | | |
| --- | --- | --- | --- | --- |
| Yadu | Turvaśu | Pūru | Druhyu | Anu |

*Figure 2.16*

The first story has two generations of descent, and the second has a successor for the primal king. Dumézil's comparative studies imply that both of these elements are required for a full statement of the basic theme, which is that of succession by the second generation (1973a: 102-4, 114-7). These two elements are found in the Irish story about the king, Eochaid Feidlech, who has four children but who is succeeded by a grandson, Lugaid, and not by one of his children. The Indian stories serve to confirm aspects of the structure of the Irish narrative and to demonstrate the Indo-European nature of their common theme, but it is to the Irish tale that we are indebted for an extraordinarily compact mythological statement which suggests a new way of looking at the origin of the cosmic structure under review.

Eochaid's four children are Nar, Bres, and Lothar, his sons, and Clothru, his daughter. Lugaid is the son of all four of Eochaid's children, his birth resulting from Clothru's having intercourse with her three brothers. This tale has its more extended counterpart in the story of Mādhavī's giving birth to a series of sons by successive fathers. In the case of Lugaid the three fathers are related to the functions not by their special virtues, as in the Indian story, but by the different body parts. Lugaid, who is known as Lugaid of the Red Stripes, has red circles round his neck and waist which mark off the three parts of his body. His head (first function) resembles that of Nar, his upper body (second function) resembles that of Bres, and his lower body (third function) resembles that of Lothar (Stokes and Windisch, ed., 1891-7: 332-3). His inheritance from his mother is no particular part but is probably to be found in the correct ordering of the three parts into the whole. The total statement seems to be one of distribution among four children followed by a recombination in a grandson, as seen in figure 2.17.

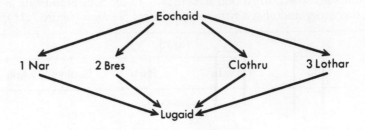

*Figure 2.17*

It is the distribution that concerns us here, and distribution of the special qualities or body parts of a primal being is a familiar Indo-European motif. Dumézil has studied, in comparison with Yayāti, the Iranian Yima whose 'Glory' left him in three parts, each of which fell into the hands of a man associated with one of the three functions (1973a: 38-54, 109-10). The cognate figure, the Scandinavian Ymir, is physically cut up by the three brothers, Odin, Vili, and Ve, and the parts are used to make the cosmic regions. It is important for the present discussion that there has been scholarly recognition of the fact that the division of the primal king can alternatively be stated in genealogical terms and that, for example, the account of the Germanic Mannus, whose three sons are the originators of three tribes, is an instance of this motif (Lincoln 1975-6). In the cases of Yima, Ymir, and Mannus, there is a division of a male figure, and it is stated or suggested that the division is into three parts. The extreme interest of the Irish story lies in the fact that Eochaid has a daughter as well as three sons, for the implication is that the primal being whose division results in three males and a female must contain a female element, that is, that the primal being is androgynous. This aspect remains latent when the distribution is cast in genealogical form but comes out very clearly as soon as it is thought of in terms of physical division.[14] The parallels quoted above seem to justify the transposition.

As the division of a male into three parts is so well authenticated, it does not seem that division of an androgyne into four parts should be put forward as a substitute for the familiar idea, but there is no need, in fact, to choose between the two ideas, for the derivation of four from an androgynous primal being is compatible with the derivation of three from a male if there are two stages in the division. As shown in figure 2.18, we can postulate (1) the division of the

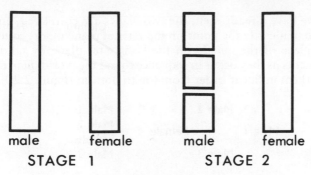

male          female          male          female
STAGE 1                        STAGE 2

*Figure 2.18*

androgyne into two parts, the male and the female;[15] and then (2) the division of the male into three parts related to the three functions. It can be seen how well this cosmogonic sequence would fit with the structural elements that have already been studied. After the splitting of the androgynous figure from top to bottom, the female, remaining entire, would provide the vertical axis – the cosmic pillar or tree – while the three parts of the male would provide the three cosmic levels (see figure 2.19). As for the location of the male and female on

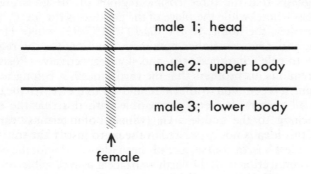

male 1; head

male 2; upper body

male 3; lower body

female

*Figure 2.19*

the horizontal plane, the sexual organs of the androgynous standing figure would correspond to the world quarters as in the Hausa instance mentioned above if the figure had the male face to the east, and this seems the most probable orientation in the Indo-European context.[16]

As we have already seen, the two planes are correlated, and it is possible to identify the point on the vertical plane that is equivalent to the place of the female on the horizontal plane by taking the components as they occur in sequence round the world quarters and setting them in linear order from top to bottom (figure 2.20).

|  |  |  |  |
|---|---|---|---|
|  | Male 2 |  | Male 1 |
| Male 1 |  | Female | Male 2 |
|  |  |  | Female |
|  | Male 3 |  | Male 3 |

*Figure 2.20*

This introduces the female between atmosphere and earth and suggests that the goddess may have a place on the earth's surface and that the 'earth' of the three Indian worlds of heaven, atmosphere, and earth should be regarded as stemming from a concept of the earth excepting its surface – the world below. The Indian emphasis favoured the upper levels of the cosmic system, but the Indo-European peoples did not exclusively worship deities of the earth's surface and upwards.[17] The position of the female on the vertical plane suggests that the three cosmic regions of the gods are the regions 'elsewhere' while the place of the goddess is the 'here' or, as we saw earlier, the home. In the *Iliad* (15.187-93) where Hades, Poseidon, and Zeus remember how they drew lots for the regions they were to rule – darkness, sea, and sky, respectively – Poseidon forcibly reminds his brothers that the earth does not belong to any one of them; it is common to all. One could say that on earth the three gods are all visitors. It seems compatible with this that the earth should belong to the goddess Ge (whose name means 'earth'), although this idea is not expressed in the passage in Homer.

The goddess has, of course, already been identified with the world tree, but a connection with the earth's surface is also possible, and Ge appears to have been envisaged both as the surface of the earth and as the world tree (just as in the present theory) by the Greek writer

Pherecydes of Syros, who lived in the sixth century BC. According to his cosmology, there was in the beginning, besides an androgynous being called Chronos and a god called Zas (= Zeus), a goddess who was called at first Chthonie but whose name later became Ge 'because Zas gave her the earth as a gift of honour,' that is, as a bridal gift. This gift is also described as an embroidered robe woven by Zas, and M.L. West, whose account of Pherecydes I follow, explains that the robe 'has the visible outer surface of the earth depicted on it' (1971: 9-12, 15-20, 59). As West says, one expects the goddess to wear the robe she has been given, and yet Pherecydes also speaks of the embroidered robe as being on a 'winged tree'; West concludes that the goddess and the tree are one. He refers to the tree as 'a self-supporting frame' that 'connected the different levels of the cosmos'. The total picture of the goddess's place is that of cosmic tree plus earth's surface.

A pattern can be discerned in the vertical division of space, which can also be discerned in the division of the year cycle, and so it is possible to offer the pattern as a general principle of cosmic structure. Only a rather complex statement can do it justice, and my tentative definition is as follows: the pattern consists of a triad of three components of the same type completed by a component of a different type which is the whole, this whole being represented by a fraction introduced into the sequence of the triad as well as being present as a totality. In the year cycle, there are three seasons related to the three gods of the functions – spring, summer, and winter. The goddess is related to the year as a whole which is represented by a short period of time that comes, I suggest, at the point between summer and winter where autumn later emerged (chapter 6; Nilsson 1920: 71-6). On the vertical plane, there are three cosmic regions related to the gods of the functions – heaven, atmosphere, and netherworld. The goddess is related to the whole (the world tree which connects the cosmic regions) and to a thin slice representing the whole which is the earth's surface.

On the horizontal plane with its four directions the part of the goddess is an entire quarter, and we saw that on this plane the goddess and the god of the third function shared a half in opposition to the gods of the first and second functions. This opposition, which is also a complementariness, has been studied by Dumézil particularly in the context of the wars of the Aesir and Vanir and the Romans and Sabines, where the first and second functions are pitted against the third function.[18] Other scholars working with different material have also demonstrated the importance of moieties among the Indo-Europeans.[19] In the cases of the vertical plane and the year cycle, the part of the goddess is a mere fraction which is probably to be regarded as negligible in size; this means that the alliance of the goddess and the god of the third function can only have a part equal to that of the alliance of the gods of the first and second functions if the section of the god of the third function is as large as the sections of the gods of the first and second functions put together. Of course, it is conceivable that the two moieties did not always match each other in size, but there is a consideration connected with the hierarchical relationship of the three functions that supports the suggestion just made. The parts of the body are not equal in size: the lower body is larger than the upper body and the upper body larger than the head. On the social level, as is made explicit in the *Republic,* the bulk of the population belongs to the third function, a smaller group belongs to the second function, and the smallest group of all belongs to the first function. Size is related inversely to value. If we apply the hierarchical model to the vertical plane and the year cycle, the part of the first function would be smallest, the part of the second function next in size, and the part of the third function largest. When the whole is divided into six equal sections with one allotted to the first function, two to the second, and three to the third, the complementary moieties receive equal shares;[20] these proportions are used in figure 2.21, which allows the suggested structures of the horizontal plane, the year cycle, and the vertical plane to be seen comparatively.

When I first suggested that the Indo-European cosmos included both triadic and quaternary types of structure, I said that 'the system which could accommodate both was one of some intricacy' (Lyle 1979). The further suggestion I am now making – that the structure was also dyadic even where the triad was concerned – is in keeping with that earlier impression. The postulated system is remarkably fluid since one aspect can be transformed effortlessly into another – three into four by the treatment of the whole as an additional component, and three into two through the treatment of two of the three parts as equivalent in size to the third.

In all this, the triad remains of key importance, and the study of cosmic structure confirms the value of Dumézil's work which has ensured that we will not lose sight of the place of the triad in Indo-European ideology. As regards Dumézil's claim that the concept of

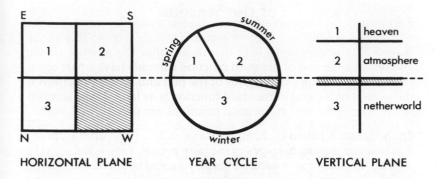

*Figure 2.21*

the three functions is exclusively Indo-European, however, there is room for doubt, for it could be that it is an archaic feature that has been more fully retained by the Indo-Europeans than by other peoples. This preliminary attempt to place it in the context of cosmic structure may well open the way to fruitful comparison with other cultures (see chapters 6-7 and 10-14).

# 3. 'King Orpheus' and the Harmony of the Seasons

In the Scottish ballad of 'King Orpheus' (Child 19),[1] after Orpheus has finished playing his pipes to the fairy people, he is asked what he would like to have as reward for his playing and he asks for his queen, Isabel, the Scottish Eurydice, and takes her safely back home. I want here to focus on the musical performance which leads to this happy outcome for it is, as we shall see, in a way comparable to the much more obviously splendid harmony of the spheres which Orpheus is described as hearing in another Scottish work, the fifteenth-century *Orpheus and Eurydice* of Robert Henryson:

> In his passage amang the planetis all,
> He herd a hevynly melody and sound,
> Passing all instrumentis musicall,
> Causid be rollyng of the speris round.[2]

Henryson is writing in the Neoplatonic tradition and his learned treatment of the theme of the harmony of the spheres falls readily into place in the line of development of classical thought in Western literature. The treatment of music in the ballad does not do so and I shall argue that it belongs to a very early conceptual framework which preceded the idea of the harmony of the spheres and was largely replaced by it in the mainstream of Western tradition. I am, I believe, making the point about the ballad music for the first time but I am not breaking new ground in claiming that the harmony of the spheres is a development from an earlier concept. M.L. West notes with approval that Walter Burkert 'has suggested, on the basis of several ancient references and parallels among other peoples, that the later theory of a harmony of the planetary spheres developed from a simpler conception of a correspondence between the four seasons and four notes or intervals.'[3] It is this 'simpler conception' that is in question here but in my view the correspondence cannot be understood in terms of the relatively late four seasons. The Indo-European year had three seasons only – summer, winter, and spring – and it is to this group of three seasons that the three types of music in 'King Orpheus' may be seen to correspond.

The different kinds of music that Orpheus plays are defined by their influence. The first kind causes 'noy' (grief) and the second 'joy' while the third has a comforting effect – it could have made a sick heart 'heal' (well). The ballad lines are:

> First he played the notes of noy,
> Then he played the notes of joy,
>
> And then he played the gaber reel,
> That might a made a sick heart heal.[4]

The romance of *King Orphius*, which is the source of the ballad, has unfortunately survived only in fragmentary form and does not include the section where Orpheus plays in the hearing of the fairy people and so it is impossible to say whether the three types of music occurred there. Later in the narrative, however, Orpheus has occasion to play the third type of music with its power of healing grief when his host, the burgess, is overcome with sorrow:

> He begouithe to weip with this
> That was the worthiest, I wis.
> Than Orphius tuik his hairp with this
> For to comfort the burgess
> And sa he did into that stound,
> The hairp it gaif sick ane sound.[5]

In this text from a sixteenth-century manuscript Orpheus plays the harp and not the pipes as in the ballad and it is in connection with the harp that the concept of the three types of music is familiar in the Celtic tradition to which the motif in the ballad apparently belongs. In Irish, the three strains are called *goltraige*, *gentraige*, and *súantraige* (music that provokes tears, music that provokes laughter, and music that induces sleep). There is a superficial difference between the sleep-inducing music here and the healing music in 'King Orpheus' but at a deeper level the difference is resolved for the type of music which took its name from the power to send to sleep was also a healing music which brought relief from suffering by means of sleep, as in the description of the effect of the music emitted by the three gold apples on the fairy branch in *The Adventure of Cormac*

*Mac Airt*: 'at that melody the men of the world would sleep, and neither sorrow nor affliction would oppress the people who hearkened to that melody' (Hull 1949: 877).

According to an Irish triad these three types of music are the 'three things that constitute a harper' (Meyer, ed., 1906: 16, no. 122) and a skilled performer was expected to excel in all three. In *The Second Battle of Moytura* when Lug claims to be a harper he is called on to demonstrate his skill and does so by playing in the three different ways:

> 'Let a harp be played for us,' say the hosts. So the warrior played a sleep-strain for the hosts and for the king the first night. He cast them into sleep from that hour to the same time on the following day. He played a wail-strain, so that they were crying and lamenting. He played a smile-strain, so that they were in merriment and joy.[6]

In several Irish stories the performance of the three types of music culminates with the musician playing the sleep music which leaves him in control of the situation. At a later point in *The Second Battle of Moytura* the Dagda, accompanied by Lug and Ogma, enters the hall of their enemies, the Fomorians, and summons his harp which is hanging on the wall:

> Then the harp went forth from the wall, and kills nine men, and came to the Dagda. And he played for them the three things whereby harpers are distinguished, to wit, sleep-strain and smile-strain and wail-strain. He played wail-strain to them, so that their tearful women wept. He played smile-strain to them, so their women and children laughed. He played sleep-strain to them, and the hosts fell asleep. Through that (sleep) the three of them escaped unhurt from the Fomorians though these desired to slay them.

In an Arabic parallel (Ibn Khallikan, trans. de Slane 1843-71: 3.309), where the musician also slips away after the sleep music has taken effect, the emphasis is on the musician's mastery and we are given a little more information about his playing. It appears from the alteration of the strings that the different kinds of music are played in different modes.

> The prince then ordered some of the most eminent performers of instrumental music to be brought in, but not one of them could touch his instrument without exciting Abū Nasr's disapprobation. 'Have you any skill in this art?' said Saif ad-Dawlat.

– 'I have,' replied the other, and drawing a case from beneath his waistband, he opened it and produced a lute. Having tuned it, he began to play and cast all the company into a fit of laughter. He then undid the strings and, having tuned it in another manner, he played again and drew tears from their eyes. Mounting it a third time, in a different key, he played and set them all asleep, even the door-keepers, on which he took the opportunity of retiring and left them in that state.

In Arabic tradition the various melodic and rhythmic modes were related to individual strings of different pitches as shown, for example, in this description of the effect of the string called *zīr* (C):

> What appears through the movements of the *zīr* string in the action of the soul, are the joyful, glorious, victorious actions, and hardness of heart and courage, and so forth. And it is related to the nature of the *mākhurī* rhythm and what resembles it, and there results from the faculty of this string and this rhythm that they strengthen the yellow bile, moving it, and silencing the phlegm, quenching it.[7]

In Ireland, too, each kind of music is found related to a single string. *The Lay of Caoilte's Urn* gives a full account of a harp with three strings each of which is associated with the effect of one of the three types of music:

> The household harp was one of three strings,
> Methinks it was a pleasant jewel:
> A string of iron, a string of noble bronze,
> And a string of entire silver.

> The names of the not heavy strings
> Were *Suantorrglés*; *Geantorrglés* the great;
> *Goltarrglés* was the other string,
> Which sends all men to crying.

> If the pure *Goiltearglés* be played
> For the heavy hosts of the earth,
> The hosts of the world without delay
> Would all be sent to constant crying.

> If the merry *Gentorrglés* be played
> For the hosts of the earth, without heavy execution,
> They would all be laughing from it,
> From the hour of the one day to the same of the next.

If the free *Suantorrglés* were played
To the hosts of the wide universe,
The men of the world, – great the wonder, –
Would fall into a long sleep.[8]

The three kinds of music, then, as played by Orpheus in the ballad, belong to a tradition where they are associated with three strings, presumably giving out notes of different pitches as in the more fully documented Arabic material. It is now possible to turn to the connection claimed by Burkert between the notes and the seasons. The three strings are not directly linked with the seasons in the Irish tales, although the Dagda's summons to his harp in *The Second Battle of Moytura* does include the words, 'Come summer, Come winter!', but they are connected quite explicitly in an account of the invention of the lyre written in Greek. According to Diodorus Siculus, Hermes 'made a lyre and gave it three strings imitating the seasons of the year; for he adopted three tones, a high, a low, and a medium; the high from the summer, the low from the winter, and the medium from the spring'.[9] Similarly Boethius, who includes autumn among the seasons, speaks of Mercury as the inventor of a lyre with four strings each of which has its equivalent season.[10]

As long as the connection is confined to that between a note and a type of music, it can be thought to depend on a musical association of some kind, but once each note is linked with a season it becomes completely clear that a system of correspondences is being referred to and such a system, by its nature, is not going to be confined to a single set but will involve a whole network of correspondences, some of which may be quite arbitrary. We are entering a different realm of thought from that of logical statement, a realm that Burkert sees as satisfying some of the same requirements as the later scientific thinking (1972: 399):

> Order and pattern ... which the human spirit craves, are to be found not only in the form of conceptual rigor and neatly logical structure, but, at an earlier level, in richness of mutual allusiveness and interconnection, where things fit together 'symbolically.'

Strangely enough, it is Plato's *Republic*, the very work which formulated the conception of the harmony of the spheres, which also gives us a particularly clear insight into the simpler conception of the harmony of the three notes, but before considering the *Republic* it will be as well to have a closer look at the three notes themselves. Part of their symbolic force seems likely to have lain simply in the ordering of high, low and intermediate pitch, but the notion of harmony comes over particularly strongly when the three notes are concordant and it seems possible that the use of three concordant notes was not confined to Greece, although the examples come from there. The Greek scale referred to is that where the note of lowest pitch (*hypate*) and the note of highest pitch (*nete*) are an octave apart while the intermediate note (*mese*) is a fourth above *hypate* and a fifth below *nete*.[11] Of the three, *mese* is in the position of control as one can see, for instance, in a reference by Plutarch to a cosmic scheme in which 'the sun himself as *mese* holds the octave together'.[12]

Turning now to Plato, we find that he expresses in the *Republic* a system of correspondences by threes on the social, psychological and physical planes. Society is divided into three groups: 1 rulers, 2 soldiers, and 3 farmers 'and the rest'. Man's soul has three divisions: 1 the rational, 2 the spirited, and 3 the appetitive; and his body has three parts: 1 head, 2 upper body, and 3 lower body. The threes correspond and have a hierarchical order which at one point Plato expresses by reference to metals of descending values.[13] These correspondences can be tabulated as in figure 3.1

| I rulers | rational | head | gold |
| 2 soldiers | spirited | upper body | silver |
| 3 farmers | appetitive | lower body | iron and bronze |

*Figure 3.1*

Plato also makes the musical comparison. In the just state, all parts are ruled by justice which assures that each part does what is fitting to it; similarly, the just man will have all parts of his soul acting in harmony. A man will be just 'when he has bound together the three principles within him, which may be compared to the higher, lower, and middle notes of the scale ...' (*Republic* 443; trans. Jowett 1888: 137). However, Plato does not say which note corresponds to which place in his hierarchical ordering and the matter gave rise to speculation among commentators. One solution of the problem was the simple and natural one of equating the two extreme notes with the extremes of the body – head and lower body – leaving *mese* to be equated with the upper body,[14] but Plutarch seems closer to Plato's thinking when he concludes that the controlling power of *mese* belongs to the rational part, the head. He argues that, although by 'local position' the head is at one extreme, the top, we should pay attention to function rather than to position:

> In fact it is incidental that in the body of man the rational part has been situated as first in local position; but the foremost and most sovereign function belongs to it as *mese* in relation to the appetitive as *hypate* and to the spirited as *nete* inasmuch as it slackens and tightens and generally makes them harmonious and concordant by removing the excess from either ... [15]

This gives the head as equivalent to the note of intermediate pitch while the upper and lower body are equivalent to the notes of extreme pitch. The correspondences, while functionally satisfying, are certainly awkward to envisage and Plato's treatment of the tripartite soul in the *Phaedrus* should probably be brought in here for it suggests a different triple division of the human frame which gives the head in the central position. In this work the rational part, the head, is viewed as a charioteer driving two horses, the white one on the right representing the spirited (which is related elsewhere to the upper body) and the black one on the left representing the appetitive (which is related elsewhere to the lower body) (*Phaedrus* 246, 253; trans. Hamilton 1973: 50-1, 61-2). Plato is glancing at a body-part correspondence which I think may prove very helpful in establishing the early system of cosmic harmony, but the present point is that it is possible to equate the right and left sides of the body with the extreme notes and the head as centre with *mese*. There is now no conflict between our understanding of function and the immediate image of the intermediate element in the middle position.

In addition to the reference to the harmony of the three notes, the *Republic* offers some comments on the musical modes and their influence in the course of the discussion of education.[16] Plato approves of only two modes, Phrygian and Dorian, which he associates with courage and moderation. As he is concerned with educating the upper levels of the hierarchical society it seems highly probable that the Phrygian mode, linked with courage, is considered appropriate to his class of soldiers while the Dorian mode, linked with moderation, is appropriate to his class of rulers. The Lydian and Ionian modes which are used for dirges and laments or are linked with drunkenness, softness and idleness, are rejected as educationally unsuitable and are apparently appropriate to the rest of the people. The 'dirges and laments' in the *Republic* are clearly equivalent to the 'notes of noy' in 'King Orpheus' and I take it that the 'notes of joy' are at the opposite extreme, while the power to allay grief found in the music of healing links this type of music to the moderation of the Dorian mode. It is now possible to suggest how the three kinds of music in 'King Orpheus' connect with the sequence of correspondences that has emerged from this discussion of Diodorus Siculus and Plato (figure 3.2).

| winter | spring | summer |
|---|---|---|
| hypate | mese | nete |
| left side of body | head | right side of body |
| lower body | head | upper body |
| appetitive | rational | spirited |
| farmers | rulers | soldiers |
| iron and bronze | gold | silver |
| Lydian and Ionian modes | Dorian mode | Phrygian mode |
| music of grief | music of healing | music of joy |

*Figure 3.2*

The correspondences in Plato have been thought to be the result of Oriental influence[17] and it is certainly possible to point to parallels in Indian sources. The three parts of the soul in Plato may be compared with the three *guṇas* or constituents of Nature in Sāṃkhya philosophy – *sattva, rajas* and *tamas* (Gonda 1976: 208). The three parts of the body distinguished in the *Śatapatha-Brāhmaṇa* are 'what is below the navel, what is above the navel and below the head, and the head itself,'[18] and in the *pravargya* ritual the three parts of the body are connected with the three classes of society – *brāhman, kṣatriya* and *vaiśya* (priest, warrior and farmer) (van Buitenen 1968: 124-5). Similarly, in the horse sacrifice the three parts of the horse's body are marked out with three different metals, including gold and silver (Dumont 1927: 277, 339). The three classes of society are linked with the three seasons (*Śatapatha-Brāhmaṇa* 2.1.3.5). The *Rigveda* 'was and is sung to three notes'.[19]

However, we find some of the same correspondences when we turn to the West. In *Celts and Aryans* (1975: 96-7) Myles Dillon, agreeing with Dumézil's finding about the tripartite division of society, connects the Irish *fili*, *flaith* and *aithech* with the Indian *brāhman*, *kṣatriya* and *vaiśya*. The triple division of the body is clearly indicated in Irish legend by the red lines round the neck and waist of Lugaid of the Red Stripes (Stokes and Windisch, ed., 1891-7: 332-3), and we have seen above that there were three harp strings, each of which was of a different metal. It may be that the Greeks were able to draw something from the common Indo-European heritage as well as borrowing specifically Eastern developments of ideas of cosmos. At any rate, the Greek interest in music has supplied a point of contact for the sequence of the three kinds of music in 'King Orpheus' and allows us to set it in a probable context where it can be seen as one of the foundation elements in a general fitting together or *harmonia* that can be expressed in terms of music or, with equal aptness, in terms of the seasons of the year.

# 4. The Circus as Cosmos

John Malalas quotes Charax of Pergamum as saying that 'the hippodrome was built according to the organisation of the cosmos,'[1] and in this chapter I explore some of the implications of this statement. The Roman circuses and the hippodromes of the Eastern Roman Empire based on the Roman model can be regarded as functioning on the level of the ideal as well as the real, but, since they had to meet the practical needs of chariot racing and its spectators, it is hardly to be expected that their structure would mirror the cosmos at every point.[2] In view of this, it has seemed advisable to consider written accounts of the circus (which did not have to keep practicalities in mind) in addition to the design of the circuses themselves. Tertullian, Cassiodorus, Isidore of Seville, the anonymous author of *De Circensibus*, Corippus, Malalas, and John Lydus all provide treatments of circus symbolism which have proved useful in this attempt to understand the circus as an expression of the cosmic plan.

## The spatio-temporal system

In a study of the theoretical writing on solar and cosmic symbolism related to the circus, Gilbert Dagron observes (1974: 333):

> ... il tend à substituer un plan idéal au plan réel du Cirque-Hippodrome. L'obélisque, considéré comme un point central, définit un espace circulaire, et c'est au cercle que renvoie aussi le 'cycle' des saisons, des mois et des signes. Or la course est tout autre chose: l'alternance de passages d'un côté, puis de l'autre côté de l'axe central qu'est la Spina avec ses bornes.

Dagron refers to Lydus, who mentions the central obelisk ('... and there is a pyramid in the middle of the stadium'),[3] and also to Corippus, whose account of the connections of the circus with the seasons and 'the circle of the full year' is especially clear and explicit:

In times of old our fathers established shows in the new circus in honour of the welcome sun. By some mode of reasoning they thought that there were four horses of the sun, signifying the four seasons of the continuous year, and in their image they established the same number of drivers, alike in meaning, number and appearance, and the same number of colours, and made two factions with opposing loyalties, as winter cold vies with the flames of summer. For green is of the spring, as the meadow, the same colour as the grass, the olive burgeoning with foliage and all the woods grow green with luxuriant leaves: red is of the summer, shining in rosy garb just as some fruits redden with glowing colour: the blue of autumn, rich with dusky purple, shows that the grapes and the olives are ripe: white, equalling the snow and the frost of winter in brightness, joins together and allies with blue. The great circus itself, like the circle of the full year, is closed into a smooth ellipse by long curves, embracing two turning posts at equal distance, and a space in the middle of the arena where the course lies open ...[4]

The turning-posts at either end of the barrier (referred to by Dagron as the spina)[5] call for special attention. According to the theoretical writing (for example, *De Circensibus,* where it is said that 'the two turning-posts indicate east and west') they are supposed, as Dagron notes, 'représenter l'Orient et l'Occident, bien que l'édifice ne soit orienté ni à Rome, ni à Constantinople,'[6] and this piece of information is useful when constructing the 'plan idéal'.

The design of the 'ideal' circus can be shown in the form which Dagron points out that the symbolism implies, a circular shape with a central point (obelisk) and a line running through this central point oriented east-west (barrier). The circle is not an unbroken and featureless ring, however, for the barrier which divides the course in two has to be shown in relation to the opening by which the chariots enter and leave the circus, and the east-west orientation of the barrier indicates that this opening is at either the east or the west. West has been preferred to east here since this direction is more in keeping with the orientation of the Circus Maximus (the prototype of the Roman circuses), where the opening is between west and north. As the circus was held to represent the cycle of the four seasons – spring, summer, autumn, and winter – it is apparent that, in idea, the circle was divided not only along the east-west axis but along the north-south axis as well, so as to yield a division into four equal parts. All these features are illustrated in figure 4.1.

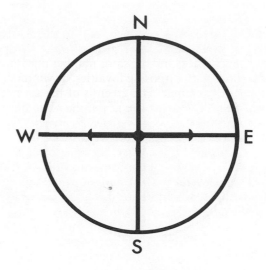

*Figure 4.1*

There are two indications as to how the particular seasons should be related to this design. The break in the circle shows where there is a beginning-and-end or point of transition and enables the first and last of the seasons to be located in the quarters next to it, while both the writers on the circus who list the seasons in sequence, Corippus and Lydus, begin with spring, which suggests that spring is the first season of the series.[7] In figure 4.2, the seasons are shown placed in order round the circle of the ideal circus, following the direction of the apparent movement of the sun.

In a total cosmic ordering, structures of varying sizes are of the same design and the cycle of the seasons will be of the same pattern as the shorter lunar cycle and the longer human life span. Philo of Alexandria notes (*De Opificio Mundi* 101, trans. Colson and Whitaker 1929-62: 1.80-3) that the period of the moon's visibility is divided into four parts just as the year is divided into four seasons, and makes an explicit comparison with a runner on the race-course:

[28] is the number which brings the moon back to her original form, as she retraces her course by lessening till she reaches the shape from which she began to make perceptible increase;

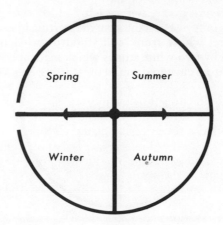

*Figure 4.2*

for she increases from her first shining as a crescent till she
becomes a half-moon in seven days, then in as many more she
becomes full-moon, and again returns the same way like a
runner in the double race-course [*diaulos*], from the full to the
half-moon in seven days as before, then from the half to the
crescent in an equal number of days: these four sets of days
complete the aforesaid number.

Philo also makes the comparison between the human life cycle and
a race on the double race-course:

The limit of man's progress upwards is his prime; when he has
arrived at this, he no longer goes forward, but, like runners in
the diaulos who turn round the post and come back on the same
course, he pays back to feeble old age what he received from
vigorous youth.[8]

Since the month, the year, and the whole of human life can be
compared to a one-lap race, the most basic expression of the
symbolism of the circus seems to require only one turning-post
placed at the far end from the opening. When the circus is seen in this
way, it is possible to relate it directly to the simple pattern of the

chariot race at the funeral games for Patroclus in the *Iliad*. The relevant features of the race are described by H.A. Harris:

> Homer gives no indication of the length of the course. It was a simple out-and-home, with a left-handed turn round a post which was only just in sight from the starting line. This post was a dead tree, whose trunk still stood to a height of six feet, supported by two white stones which made it conspicuous. ... The spectators clustered round the starting line which was also the finish.[9]

The widdershins direction of racing indicated by the left-handed turn round the post was also the practice of the Roman circus; it is perhaps to be understood as a reversal, associated with death (Frazer 1911: 92-3). The structure of the Homeric race is that of beginning and end at the same place and of middle at a turning point distant from the beginning and end, as shown in figure 4.3.

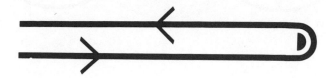

*Figure 4.3*

Quite evidently, this race is dual, and Dagron noted that there was a duality in the circus where the course was divided into halves by the barrier. He found this completely different from the circle including the seasons, but it is apparent from Philo of Alexandria's treatment that the lunar cycle (and so, by analogy, the year cycle) could be seen as falling into two main parts divided at the turning-post (equivalent to full moon) while at the same time falling into smaller divisions. For Philo of Alexandria, a Jew, it was natural to divide the cycle of the visible moon into four periods of seven days, two of which related to the increase of the lunar orb and two to its decrease. The Roman month has three sections not four, but in spite of this it provides a dual division just as the model based on the seven-day week does. The Kalends on the 1st corresponds in idea to new moon and the Ides on the 13th or 15th to full moon, so that the threefold division coexists with a dual division which contrasts the waxing and waning halves of the month.[10] Figure 4.4 shows the lunar cycle in relation to the circus in terms of **a** the account by Philo of Alexandria, and **b** the divisions of the Roman month.

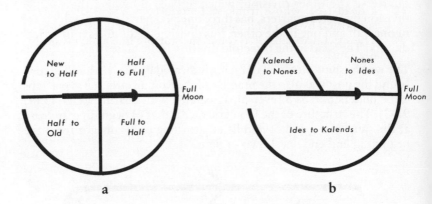

*Figure 4.4*

## Tetrad and triad

I believe that the road to understanding the archaic substratum of the circus symbolism has been blocked by the lack of a theory explaining the relationship within a system of meaning of a four-part division and a three-part division like the ones shown here. The seven-day week which fits a quadripartite division of the month was imported into Rome from the east, perhaps in the Augustan period,[11] and the four-season year is no more fundamental, for it was preceded by a three-season year consisting of spring, summer, and winter.[12] In view of this, it is not surprising that the symbolic connections of the circus with four weeks and four seasons should be dismissed as late and derivative.[13] However, the statements by the writers on symbolism concerning the four divisions of time can have validity if, as I suggest, their symbolic meanings evolved from an earlier structure where three divisions of time stood in a fixed relationship to a companion set of four, so that each part had its known equivalent in the four-part division. In that case, the symbolic associations of the four divisions of time could apply only in modified form to the three divisions of time but could apply directly to the related fourfold division.

Since I have discussed in detail in chapter 2 my theory about how the fourfold and threefold divisions operated within a total system, I shall merely summarise some of my main conclusions here. The division into four equal parts belongs primarily to the horizontal plane; it yields a symmetrical duality with each half containing two quarters. The other division, which applies elsewhere than in the division by world quarters, has three unequal parts, two of which are in one half and one in the other; this division yields an asymmetrical duality. The triad in the threefold division is the male triad, and each of the sections corresponds to one of the world quarters. The world quarter that does not correspond to a section of the male triad is female, and its equivalent in the asymmetrical division is 'the whole', which, though interfused through the three sections, may be represented by a single point.

As all aspects of the triad and tetrad are in correspondence with each other, it would have been simplicity itself to transfer from a three-season year and three-part month to a system of four seasons and four weeks; all that had to be done was to apply the world-quarter model and expand the triad to a tetrad by treating the female whole as a part. The resulting scheme, based throughout on four, is very straightforward but is flat and unilluminating, and cannot have been the system of the earlier time when the four-season year and four-week month were not current in the culture. It seems to me that, if the operation of the circus on the plane of ideas is to be understood, an attempt has to be made to tackle the greater complexity of a system which converted readily between four and three, employing a correspondence between the fourfold division of the world quarters and the triad plus the whole. Naturally, the proposed cosmic scheme can only be hypothetical at present, but the fresh approach which I use here does offer for examination a coherent system that accords with a number of indications in Latin and Greek sources and should supply a basis for future discussion.

## The colours

The association of the four factions with colours is a striking feature of the Roman circus. I have not dealt with the matter of colour up to now for, although it can be said immediately that it is probable that the four colours – green, blue, red, and white – would, as in other cosmic schemes, be located in the world quarters, the vital matter of their order cannot be settled without bringing broad considerations to the interpretation of the Roman evidence.

Recent studies of colour terminology indicate that there is a progression from distinguishing two colours, 'white' and 'black' (or light and dark) to distinguishing three colours, 'white', 'red', and 'black' (or light, dazzling, and dark) (Berlin and Kay 1969; Rowe 1972). The fourth colour term to enter the terminology refers to either green or yellow, and the fifth names the one of the two colours, green or yellow, that has not already occurred. Blue, as a distinct colour term, appears late, coming after both yellow and green, but if, as I suggest, the circus colours form part of an archaic system of meaning, it is likely that blue here represents black in the basic triad, as it does in the series of white and red and dark blue or black that is associated with the tripartite division of Indo-European society. The correspondences worked out by Dumézil suggest that the colours have a particular sequence, for the three colours, white, red, and blue/black, that are attached to the social groups of priests, warriors, and cultivators, necessarily occur in this order, which is fixed by the position of each group in the hierarchy (Dumézil 1958: 25-7). A valuable confirmation of this colour sequence, expressed in terms of relationship to time, occurs in Hyginus, in the story of the calf born to Minos and Pasiphaë which changes colour daily from white in the morning to red in the middle of the day and to black in the evening, and which is compared to the mulberry which ripens from white to red to black.[14]

White, red, and black form the male triad and no fourth colour is absolutely required since the female, being the whole, can be represented as combining all three colours. It seems quite possible that the archaic structure being discussed here was established at a time when no fourth colour term was available, but, with the extension of the colour terminology, the fourth term, green or yellow, would be likely to enter the system and be applied to the female. In the case of the Roman circus, the fourth colour in addition to white, red, and blue, is green, and there is actual evidence of the association of this colour with the female (the goddess, Demeter; see below).[15] In the case of Indian cosmology, the fourth colour is yellow, and the colours as they occur in order round the world quarters are white, red, yellow, and blue-black (Beck 1969: 559). When green in the Roman system is represented at the same point in the sequence, the colours as they occur round the world quarters are white, red, green, and blue. The location of the green colour suggested by the Indian parallel is in my

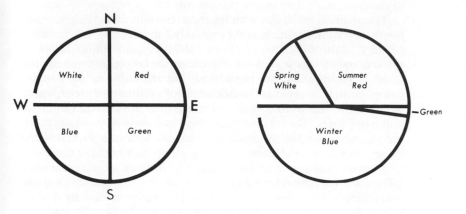

*Figure 4.5*

view supported by the correspondence between green and the earth in the writings on circus symbolism,[16] including the passage from Malalas referred to below, for I find that the earth has the equivalent place in the sequence of cosmic levels (see chapter 2).

In the asymmetrical division into three parts, the female is the whole, and the whole is represented in the sequence at the point equivalent to the green world quarter. In the case of the lunar cycle, this point comes at full moon, which has a natural appropriateness as an expression of the whole. In the circus, as shown in figure 4.4, the full moon is in the position of the turning-post distant from the entrance, and it is worth pausing to consider the symbolism of the turning-post in relation to the female. Each turning-post consisted of three cones set on a base which Cassiodorus compares to the decani of the zodiac: 'The goals themselves have, like the decani of the Zodiac, each three pinnacles.'[17] Cassiodorus is referring to the division of a thirty-day period into three equal parts of ten days, each of which is marked by the rising of a star, but, although the symbolism is expressed in astrological terms, it can be seen that, at a more basic level, the triple turning-post can be regarded as a symbol

drawing together three parts of a month period, and this is compatible with the theory that it represents the female whole which relates
to the three male sections of the month.

The position arrived at with regard to the four world quarters and
the three seasons relative to the colours of the circus is illustrated in
figure 4.5. The writers on symbolism do not equate the colours with
the seasons in this way, with the exception of red = summer, but
instead employ the correspondences that are found in the passage
from Corippus quoted at the beginning of this chapter according to
which green relates to spring, red to summer, blue to autumn, and
white to winter. These correspondences evidently rest on the resem-
blances mentioned by the writers. Tertullian, for example, states that
'White was sacred to Winter, for the gleaming white of the snow, red
to Summer because of the sun's redness,' and Isidore of Seville says:
'[They wanted] the whites [to run] for winter, because it is icy and
everything grows white with frosts; and the greens with their verdant
colour for spring, since then the vine foliage thickens.'[18] Since the
correspondences lack deep roots in a total cosmic scheme, it has not
been possible to use the explicit statements of the writers on symbolism to establish the relationship between the colours and the seasons.
Instead, this relationship has had to be established indirectly by
building up a probable sequence of the colours through their other
associations and then matching it with the known sequence of the
seasons, as shown above in figure 4.5.

*Cross-cut dual organisation*

The cosmic scheme falls into four quarters represented by the four
colours, but the quarters tend to form pairs and so the organisation
of the structure can be regarded as dual. In a simple dual structure the
same two halves would always be opposed, but the circus symbolism
offers the more complex picture of a cross-cut dual organisation
where the two halves can be formed in different ways.[19]

The dual division marked by the barrier is the powerful opposition
of the waxing and waning halves of the moon, but it is striking that
the clearest indications of contest between two halves stress the
alternative duality which can be described as that of ends against the
middle, with beginning and end (white and blue) matched against the
two middle sections (red and green). Although all possible pairings
were in operation in circus practice, Cameron (1976: 61-73) has
shown that in Constantinople it was recognised that the main

pairings were dominant blue along with subsidiary white against dominant green with subsidiary red, and that references to the opposition of the blues and greens implied the silent presence of the subsidiary partners. That the ends were conceived of as the blue half and the middle portions as the green half is indicated in terms of the lunar cycle by two Byzantine fragments concerning astrological influences on chariot racing, one of which states that the conjunction of the sun and moon is in correspondence with the blues, and the full moon with the greens, while the other says more explicitly that the moon favours the blues 'at the beginnings and the ends' – a period that is defined as from the half moon on the twenty-first, through the conjunction with the sun, and until the seventh day – and that after the seventh day the moon favours the greens.[20] This makes it completely clear that the first and last of the moon's quarters composed the half of the blues while the second and third quarters composed the half of the greens.

The discussion of the circus by Malalas includes an account of an annual contest undertaken by Oenomaus, King of Pisa, which shows blue and green in opposition.[21] One contestant wore blue and was associated with Poseidon, while the other wore green and was associated with Demeter. The result of the chariot race was held to affect the whole people, who were divided into moieties according to whether they got their living by the land (green) or by the sea (blue). Although the contest was between blue and green only, it has been seen that these colours, when regarded as representing halves, subsumed their partners, white and red. This suggests a new interpretation of Tertullian's statement about the two colours worn by the charioteers at the beginning of chariot racing in Rome (*De Spectaculis* 9, trans. Glover 1931: 256-7): 'For at first there were but two colours, white and red.' When it is remembered that, in the later period, two colours subsumed the other two in the set of four, the question arises whether, in the earlier period referred to here, blue and green might have been subsumed under white and red. It is possible that what lay behind Tertullian's statement was a situation where there were in fact four colours present in the symbolism but only two chariots took part in racing, perhaps in a ritual race settling the fortunes of moieties, like that in Malalas. If it is assumed that this was so, the point of special interest is that, given the partnerships of blue and white and of green and red, the moieties are the same in Tertullian and Malalas. The selection of different colours to represent the pairs suggests that there was a switch between the dominant and subsidiary roles of the colours in either half at some time before the early Principate, when blue and green were already pre-eminent.

It is necessary to consider also a passage (*De Mensibus*, ed. Wuensch 1898: 4.30) where Lydus states that three colours – white, red, and green – were initially involved in the chariot racing at Rome:

> By the obelisks is indicated that they were holding the contest in honour of those dying for their country, and the number of chariots contending on the occasion of the chariot-race was three and not four: the *russati* or reds, the *albati* or whites, the *virides* or flowery, nowadays called greens. And the reds claimed to belong to Ares, the whites to Zeus, and the flowery ones to Aphrodite.

It may be relevant in the context of this discussion that green is distinguished from the other colours in two ways. Firstly, Lydus, like Malalas, associates green with the female, connecting it with the goddess, Venus, while connecting the other colours with gods. Secondly, Lydus notes that green, or, as he calls it, the 'flowery' colour, has a special link with the city of Rome under its name of Flora:

> And because of the four elements they made the contests (sic) four in number: the flowery colour representing fire (sic), in honour of Rome – they used to call Rome 'Flora', as we call it 'Anthusa' [flowery] – ...

He adds that it was thought unlucky if it was defeated in a chariot race since it represented the city itself: 'And so they used to regard it as an omen of ill-fortune if the flowery colour was beaten, as though Rome herself had been defeated.' This suggests the possibility that the racing involving three colours might have been a contest between moieties represented by white and red, while the green chariot associated with the female represented the whole, so that an augury for the entire city could be drawn from its position at the finish.

At any rate, whether or not Lydus's account can be regarded as compatible with a dual scheme, the evidence of the other two writers points to the dual structuring illustrated in figure 4.6. This shows the opposed halves on either side of the heavy line, with white and red dominant in **a** (cf. Tertullian) and blue and green dominant in **b** (cf. Malalas), while the participation of green along with white and red (cf. Lydus) is indicated by the shading in **a**.

As noted above, these contending halves, consisting respectively

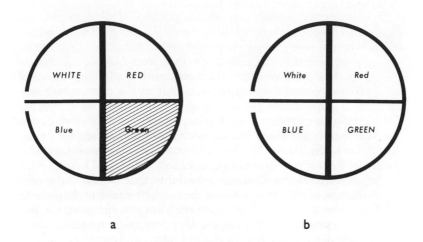

*Figure 4.6*

of white and blue and of red and green, stand in contrast to the halves divided by the barrier, where the pairings are white and red against blue and green. Although the expression 'cross-cut dual organisation' that I have applied to this feature may sound strange in the context of Roman culture, there is a hint of the presence there of this mode of organisation in the familiar division into two pairs of halves – right half and left half, front half and back half – that occurs in the *templum* in augury and surveying (Dilke 1971: 32-3), and it is possible that study of the circus in comparison with the augural *templum* will lead to fuller understanding of the symbolic nature of both these structures as reflections of the cosmos.

# 5. The Dark Days and the Light Month

I should like to offer here a fresh interpretation of the twelve days in the year that occur about the midwinter period and are sometimes identified as the Twelve Days of Christmas. They were discussed at length by Frazer who stated in conclusion (1913: 342):

> Thus we infer with some probability that the sacred Twelve Days or Nights at midwinter derive their peculiar character in popular custom and superstition from the circumstance that they were originally an intercalary period inserted annually at the end of a lunar year of three hundred and fifty-four days for the purpose of equating it to a solar year reckoned at three hundred and sixty-six days.

The actual calendars of Indo-European peoples do show intercalary additions to the lunar year but these take the form of intercalary months. Frazer was aware of this, saying that 'there are grounds for thinking that at a very early time the Aryan peoples sought to correct their lunar year, not by inserting twelve supplementary days every year, but by allowing the annual deficiency to accumulate for several years and then supplying it by a whole intercalary month'. He expressed his puzzlement over this in connection with the Celts of Gaul, commenting (342-3): 'Why they abandoned the simple and obvious expedient of annually intercalating twelve days, and adopted instead the more recondite system of intercalating a month of thirty days every two and a half years, is not plain.' The answer I have to offer is that the Indo-Europeans never did employ a twelve-day intercalary period but kept the months of the lunar year in line with the natural seasons by inserting intercalary months when required, and that the place of the twelve days in the calendar was not outside the lunar year but inside it. As I see it, the possibility of having a group of twelve days that stand in a functional relationship to the rest of the year arises when the month used in the calendar is not the synodic month of twenty-nine or thirty days but the shorter light month, the period of the moon's visibility.

Ten years before Frazer published his remarks, W.H. Roscher had studied the matter of month length in connection with Greek weeks and festivals and had concluded that months of two different lengths – 27-8 days and 29-30 days – were known in antiquity, but that the use of the short month of 27-8 days was the older (Roscher 1903: 5-7, 68-71). Frazer made no mention of this finding in his discussion of the twelve days, and yet the two ideas of the light month and a set of special days can be seen as complementary and I shall consider them together here.

The longest period of the moon's visibility is twenty-eight days[1] and the moon goes through its cycle from new moon, through full moon, to the last crescent of the waning moon in this time. As Plutarch expressed it (Griffiths, ed., 1970: 185), referring to the number twenty-eight: 'Such is the number of the moon's illuminations and in so many days does it revolve through its own cycle.' Between each pair of light months there is a gap which was called by the Romans the 'interlunium' or 'intermenstruum,' that is, the interval between the moons or the months (Roscher 1903: 6). The treatment of this time of the dark of the moon in the modern period can be seen in the Malagasy calendar in Madagascar where each month begins with the new moon and lasts for twenty-eight days. Between each pair of months there is an interval of a day or two days which is not reckoned as occurring in any month (Ruud 1960: 28-30).

The distinction that is made between light and darkness when the light month is used is a fundamental one. Nilsson in *Primitive Time-Reckoning* remarks in connection with the month, 'Apparently only the time during which the moon is visible is at first counted,' and similarly, in connection with the diurnal period, he comments that daytime is divided up while the sleeping-time at night 'has no separate parts' (Nilsson 1920: 149, 17). Light is treated separately from darkness and, as Eliade among others has noted, light is equated with life and darkness with death (Eliade 1955: 86-8).

Twentieth-century scholarship has brought about a sharpening of awareness of the importance of opposites and it is perhaps now more possible than it was in Frazer's day to feel the necessity of the tie between the polarities of light/life and darkness/death. Probably this kind of awareness of polarity should be brought to bear on the question of the year calendar, for the dismaying fact of nature is that the year has no period of darkness (for the dull days of winter are far from being absolutely dark)[2] and the deficiency, if it is felt to be such, can be made good only by cultural means. In the calendar with its twelve special days I think it is possible to trace the establishing of a duality of light and darkness within the year. If the days of the dark of the moon are notionally removed from their position between the light months so that the twenty-eight-day months run on continuously,[3] the eighteen additional days that are not in any month are freed for distribution elsewhere. Six of them are of no concern in this discussion, but I suggest in a fuller study of calendar in the next chapter that they are festival days occurring at various points in the year. The remaining twelve, however, according to the present theory, retain the connotation of darkness from their identity as the dark days between the months and, grouped together, form a period fit for the celebration of death, chaos, and reversal.[4] They are symbolically the night of the year, and this accords very well with what is already known about the twelve days. The main advance offered by the new theory is that it makes sense of the associations of the set of days in terms of calendrical structure. It also permits a fuller and more exact exploration than before of the whole range of concepts related to darkness, whether these are found in connection with night or in connection with the twelve days. The idea that darkness is a prelude to light, for example, is reinforced when one observes not only that night was considered by the Indo-Europeans to come before day (Schrader and Nehring 1917-29: 2.505), but also that the twelve 'dark' days very evidently preceded the rest of the year since it was held possible on each of them to foretell events of each of the months to come (Frazer 1913: 316, 322-5).

In historical calendars, the twelve days occur in one of the months of the year or in two consecutive months (December and/or January), and there is a certain awkwardness about this since they relate to months as entities apart from them and yet occur within the series of the months, but a passage in the *Rigveda* where the deities of the seasons are said to rest for twelve days implies the existence at one time of a system according to which the twelve-day period was outside the months,[5] and this arrangement permits a clear and direct connection between each of the twelve days and its equivalent month. When the months are regarded as necessarily synodic, then twelve months fill the lunar year, and the only place for the twelve days outside the months is outside the lunar year as an intercalary period, and it was Frazer's theory that they had this position. If the months are light months, however, they leave space inside the lunar year and make it possible for the twelve days and the twelve months to provide within the structure of the lunar-year calendar the meaningful opposition of darkness and light that has been proposed here.

# 6. Archaic Calendar Structure approached through the Principle of Isomorphism

If, as seems likely, it was a central purpose of the archaic calendar to organise and hold in place a network of meaning, it is not surprising to find that approaches made to the study of calendars through astronomy alone fail to give a complete picture of the processes involved not only in their use but also in their construction. An approach through semiotics gives promise of filling out the information to be drawn from astronomy but, of course, there is no conflict between the two disciplines for calendars, whatever their component of meaning, all work with the same underlying features: the period of night and day, the period of the month and the period of the year. The earliest calendar year was not the solar year of $365\frac{1}{4}$ days, but the lunar year consisting of twelve lunar cycles, to which an intercalary month was generally added every two or three years to keep the months in line with the natural seasons. The number of days in a lunar cycle varies between twenty-nine and thirty but is on average twenty-nine and a half and so the lunar year is normally 354 days.[1] With this 354-day period as a base, it is possible to turn to what can be learnt by taking a semiotic point of view – in particular, by applying the principle of isomorphism mentioned by Lotman when speaking of the paradigmatic type of organisation found in archaic cultures:

> The complete world picture is presented as a fixed paradigm of which the elements occur on different levels. ... The different levels are reciprocally isomorphic. (1968: 24-5)

The principle of isomorphism is not sufficient on its own,however; it is also necessary to take account of some relevant mythological material and of the archaic manner of counting in which the whole is added to the parts, as in the example relating to a group of four objects: one, two, three, four, and the group is the fifth (chapter 1; Rees and Rees 1961: 201-204; Gonda 1976: 8, 116). In the case of the calendar, the days of special significance which were celebrated as festival days appear to be days that sum up a period of time and so are wholes rather than ordinary units.

As I see it, historically known calendars contain a number of features derived from an archaic calendar where meaning was more directly evident than it is in the historical calendars which have been subjected to adaptations of various kinds, and I am offering a hypothetical reconstruction of an archaic calendar lying behind the calendars of two peoples who made a tripartite division of the year – the Egyptians and the Indo-Europeans. In the course of working with the idea of calendar, it has come to seem to me that the difference between calendars with a three-part structure and those with a four-part structure is a fundamental one, and that the difference lies in their treatment of the period of darkness.The one with the three-part structure makes an initial separation of darkness from light and proceeds with the organisation of the period of light, while the one with the four-part structure, familiar in modern times, includes the period of darkness in a continuum where it is organised along with the period of light.It is clear that the type of calendar with a three-part structure is an archaic form, and I have confined my attention to it here.

My study is divided into three sections which focus on: (1) the division into periods of darkness and light; (2) the organisation of the period of light; and (3) the alternation of the periods of darkness and light. As the principle of isomorphism is brought to bear on these different features, the structure of the archaic calendar is gradually built up.

*Section 1: The isomorphism (either natural or artificially created) between the three time periods of day, month, and year, when considered as divided into periods of darkness and light.*

Except in the polar regions, the year has no period of darkness, but the lunar cycle, like the twenty-four-hour day, divides naturally into periods of light and darkness – the light month,when the moon is visible, and the interlunium, when it is not. The longest possible period of the moon's visibility is twenty-eight days and this can be reckoned a standard or schematic light month just as the commonly occurring thirty-day month is a standard or schematic synodic month based on the longest possible lunar cycle (Roscher 1903: 5-7, 68-71). It follows from the length of the light month that in a calendar based on it there are some days in the year that are not contained in any month, a point that can be established by a glance at the Malagasy calendar used in Madagascar. This calendar is a mixture of the observational and the schematic. A month begins with the new moon and twenty-eight days of divinatory significance are counted from that point. There are then one or two extra days before the appearance of the new moon signals the beginning of the next month. This system results in there being eighteen days outside the months which are, as Ginzel puts it, 'scattered through the year' (Ginzel 1906-14: 2.133; Ruud 1960: 28-30).

I shall return to this feature of the light-month calendar shortly, but should like first to take a general look at the Egyptian civil calendar which is held to have been established some time in the three centuries between 3000 and 2700 BC.[2] Although this early calendar is not for a lunar year but instead for an approximate solar year of 365 days, I suggest that its structural features give valuable clues to the structure of the archaic lunar-year calendar even though the actual periods of time within the structure naturally have to be different. The Egyptian calendar has twelve months of thirty days divided into three ten-day periods or decads which are referred to as the decads of the beginning, the middle, and the end. The months form three seasons with four months in each, so that the year, like the month, is divided into three equal parts. Before the seasons are five days outside the months called the epagomenal days which are referred to as the birthdays of the gods, Osiris, Horus, Seth,Isis, and Nephthys. The grouping of the days of the approximate solar year as, firstly, a set of days outside the months, and then the twelve months, can be shown as: 5 + (30 x 12) = 365.

The five days act as an intercalary period bringing the 360 days of the twelve thirty-day months up to the period of the approximate solar year; it has been thought that they were an arbitrary addition to the calendar and even that their special connection to the gods was a calculated move to make them acceptable to the people (Griffiths, ed., 1970: 294-6). Ail this is speculation only. There is no firm evidence that the Egyptians ever had a 360-day calendar without the five epagomenal days and, if these days are seen, not as a late addition, but as an integral part of the structure, it becomes quite probable that the Egyptian calendar with days outside the months was modelled on a lunar-year calendar which had the same characteristic, i.e. a calendar based on the light month. That the Egyptians did at one time give a place to the twenty-eight-day month is suggested by Plutarch's references to this period in *On Isis and Osiris* where he remarks, for example, that the life or reign of Osiris was said to have lasted twenty-eight years 'for such is the number of the moon's illuminations and in so many days does it revolve through its own cycle' (Griffiths, ed., 1970: 184-7, 218-9). The importance of being able to relate the Egyptian calendar to a light-month calendar lies in the fact that the Egyptian calendar, unlike the type of calendar illustrated by the example from Madagascar, was fully schematic with the months running on independently of the lunar cycles. Such a notion exposes the possibility that a light-month calendar of the same type existed during prehistoric times. If the design of the Egyptian calendar is applied to the case of a lunar-year calendar employing a twenty-eight-day month, the result is a set of months preceded by a group of days outside the months as follows: $18 + (28 \times 12) = 354$.

This formulation brings out the relationship between the set of days outside the months, and the days occurring in ones and twos in the Malagasy calendar but there is no reason to suppose that it was that of any actual calendar. Instead, it seems likely that the highly meaningful block of twelve days for which there is Indo-European evidence was made up from part of the fund of eighteen days released by the use of the light month and that the twelve months in the calendar were preceded by a set of twelve days. In historical Indo-European calendars the rather unaccountable twelve days associated with darkness and reversal were included in one (or two) of the months, but it seems, from a single reference in the *Rigveda*, that they had once been regarded as standing outside the seasons, and this concept fits with the structural relationship in which they stood to the rest of the year, as forerunners of the months (Frazer 1913: 322-45; Eliade 1955: 62-9). It was believed to be possible, during the twelve-day period, to foretell events of each of the months of the coming year by the events on the equivalent day, and so each of the days evidently had an intimate connection with one of the months. An explanation of this close tie may be that each of the days was regarded as a dark day removed from a lunar cycle and as the complement of its equivalent light month. What appears to have happened is that the calendar year has been artificially shaped on the model of the lunar cycle. Before its light period consisting of the twelve months, there is a brief dark period comparable to the brief period of darkness before the appearance of the crescent of the new moon. A totally arbitrary 'darkness' is attributed to a block of days in the year. In the process of building the structure of the calendar year, the dark days of the lunar cycles have been gathered together and each month is a light month only, with darkness removed from it to its special position in the year. This transforms the isomorphism that had existed naturally between the day and the lunar cycle to an isomorphism between the day and the year.

The division of darkness and light within the lunar year can be shown as: $12 + (28 \times 12) = 348$ days. There are another six days in the lunar year and six days are required if each of the six 'components' of the year – the five sections (dark period, light period, first season, second season, third season) and the year itself – is bound by a day which concentrates its meaning. This archaic way of binding the units by the whole is probably most easily grasped in the context of the structure of the calendar, since festivals are widely found, and the idea of a high point within a period of time is a familiar one. Festival days often initiate the period to which they refer but this is not necessarily the case and it will be better to defer for the moment the question of the position of each of them relative to the period it concerns. In spite of this, it is possible to summarise the hypothesis arrived at in the course of this discussion as follows:

> The archaic lunar-year calendar consisted of a light period of twelve months of twenty-eight days divided into three seasons preceded by a dark period of twelve days, each of which was the complement of one of the months, and a further six days at various points in the year which were festival days relating to the dark period, the light period, the first season, the second season, the third season, and the year.

*Section 2: The isomorphism of the light period with a combination of (a) the visible phases of the moon, and (b) the vertical human body.*

It is possible to study the structure of the light period in general since the same structure is replicated in the three different time periods. This isomorphism is evident in the Egyptian calendar where the division in each case is into three equal parts, the twelve months of the year being divided into three seasons of four months, the thirty days of the month into three ten-day periods, and the twelve daytime hours into three four-hour watches.

When it comes to dividing the light month it is necessary to re-examine the matter of festival days, for the month, like the year, appears to have been subject to the grouping of units under wholes. In the way that the year has three seasons, it has three sections with their binder days but, unlike the year, it is entirely a light period that does not have to distinguish light and darkness and therefore has only one further binder day, which relates to the whole of the month. Four of the days of the month, then, are special festival days, and only twenty-four out of the twenty-eight are ordinary units to be divided into sections. This makes correspondence between the days of the month and the months of the year a very simple matter, since two days of the month are equivalent to one month of the year.

A regular division like that of the Egyptian calendar applied to the period of the light month excluding its festival days would result in three eight-day periods within the month, but the evidence from the Indo-European culture does not support this division and it seems that the answer is not so simple. In fact, the Indo-European evidence is distinctly confused, and, through this confusion, it may be possible to arrive at the probable archaic structure of the light period – just as it was possible to advance a hypothesis about the division of light and darkness on the basis of the anomalous twelve days which did not have a comprehensible place in historical calendars. What has seemed curious in the past is that there are indications of the division of the year and of the month into two parts as well as into three parts. This has been explained in terms of chronology: first there was a division into halves and then this was superseded by a division into three. Harrison, for example, commented, 'After the simple seasonal year with its two divisions came the Moon-Year with three' (Harrison 1927: 189; see also Schrader and Nehring 1917-29: 1.529-32, 2.75-6). However, in one of the examples of the three-part division, the Roman month, there is no contradiction between division into halves and division into three parts for two of the three parts fit into the first half of the month. The dividing days are the Kalends on the 1st, equivalent to new moon, and the Ides on the 13th or 15th, equivalent to full moon, and between them the Nones on the 5th or 7th (Michels 1967: 18-22, 130-2). Since this indicates both a division into halves and a three-part division, it illustrates the point that the two divisions can be contemporaneous. It seems that the concepts of beginning, middle, and end, found in the regular divisions of the Egyptian calendar, could have been present for the Proto-Indo-Europeans in an asymmetrical system which included division into halves as well as a tripartite sectioning. It was an asymmetrical system of this kind that emerged when I was engaged in study of other aspects of the total cosmic scheme apart from the calendar, and the Roman month structure, which is puzzling when seen only in terms of the connection with the moon(for there is no obvious reason for dividing the first half of the month at the moon's first quarter without making a parallel division of the second half), makes sense when put in a broader context which includes the structure of society and the microcosm of the human body.

Dumézil made a special study of the tripartite division of Indo-European society into priests, warriors, and cultivators and noted the correspondence sometimes found with the three parts of the body: head, body from neck to waist, and body below the waist (Dumézil 1973a: 105-6). While he does not believe this to be a fixed part of a cosmic framework, I argue that it is, and that the two upper parts of the body and the priests and warriors form one half of a dual division of which the other half contains the lower body and the cultivators (chapter 2). In an Iranian story much like that of Noah's ark in which Yima is told to preserve samples of all beings in a subterranean enclosure (Boyce 1975: 94-5), there is a passage which indicates, according to Benveniste's interpretation, that the place of the priests should be three measures, that of the warriors six measures, and that of the cultivators nine measures (1932: 119-21). Taking these figures down to the smallest whole numbers, the entire area can be reckoned as containing six parts of which the priests have one, the warriors two, and the cultivators three. Of course, a six-section division is well suited to fit with a calendar that has twelve months in the year and twenty-four ordinary days in the month, and the parts corresponding to the three groups in society are, respectively, of two, four, and six months, and of four, eight, and twelve days. The body does not yield exact figures but it does correspond roughly to this division, with the head as smallest part, the upper body as part of intermediate size, and the lower body as largest part. In the case of an Irish legendary figure, Lugaid of the Red Stripes, the divisions of the body are actually marked off by red lines round neck and waist (Dumézil 1973a: 105), and it is these key points as well as the main points in the course of the changing phases of the moon (Nilsson 1920a: 155, 329) that appear to be employed in the division of the light period. Both the visible cycle of the moon and the body have approximate half-way points occurring respectively at the full moon and the waist, and the crescents of the new and old moon mark beginning and end, and are equivalent to the top and bottom of the human figure. The neck, however, has no obvious equivalent in the lunar cycle (figure 6.1). Reciprocal isomorphism can really be seen at work here, for while the division of the first half into a small and larger part is on the model of the body, the representation of the whole at an approximate midpoint is in keeping with the fact that the whole of the moon's disk is visible at that time.

|  |  |  |
|---:|:---:|:---|
|  | 1 | new moon |
|  | 2 |  |
|  | 3 |  |
|  | 4 |  |
|  | 5 |  |
| neck | 6 |  |
|  | 7 |  |
|  | 8 |  |
|  | 9 |  |
|  | 10 |  |
|  | 11 |  |
|  | 12 |  |
|  | 13 |  |
|  | 14 |  |
| waist | 15 | full moon |
|  | 16 |  |
|  | 17 |  |
|  | 18 |  |
|  | 19 |  |
|  | 20 |  |
|  | 21 |  |
|  | 22 |  |
|  | 23 |  |
|  | 24 |  |
|  | 25 |  |
|  | 26 |  |
|  | 27 |  |
|  | 28 | old moon |

*Figure 6.1*
*Days of the light month and festival days corresponding*
*to the main points of the lunar cycle and the body*

The month is divided into halves of fourteen days each, with periods of five and nine days in the first half, and the day of the whole plus a period of thirteen days in the second half, i.e. 5 + 9 / 1 + 13, and it is interesting to see how well this suggested composition of the month fits with a general concept of number among the Celts which is discussed by Rees and Rees (1961:194-5, 200-1). They list a series of numbers made up on the pattern of an even number completed by the odd one above which serves as 'unifying principle' and the first three in the series are: 4 + 1 = 5; 8 + 1 = 9; and 12 + 1 = 13. There is progression from small/short to large/long with a term of intermediate size/duration in between. An asymmetrical system, unlike a symmetrical one, is capable of conveying these distinctions in addition to giving the order of beginning, middle, and end. Of course, in the asymmetrical system the middle term does not come at the midpoint, and, although the word 'middle' is a natural and convenient one, it has to be understood as that which is intermediate between two extremes and not as that which is midway.

The festival day of the beginning period falls at the beginning of the month and it seems that the transition to the middle period is

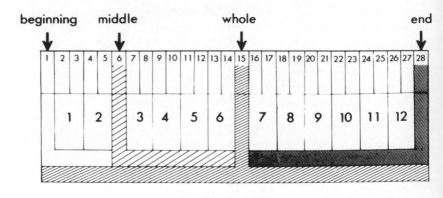

*Figure 6.2*

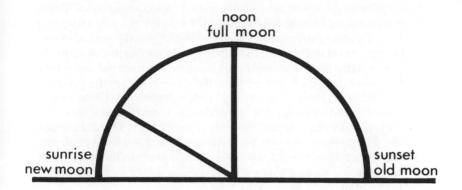

*Figure 6.3*

marked by the occurrence of its festival day at its beginning. The middle period is separated from the end period by the festival day of the whole and so the transition is clear, and the end period has its festival day at its end. Within the year, the seasons of beginning, middle, and end are spring, summer, and winter, and festival days at equivalent points to those given for the month would fall at the beginnings of spring and summer, between summer and winter, and at the end of winter.

Figure 6.2 shows the four festival days of the beginning, the middle, the whole, and the end as they occur in the sequence of the days of the month (upper row of figures) and are interspersed among the months of the year (lower row of figures). The division of the periods of the daytime can be shown most clearly in a diagram using a semicircle which conveys directly the ideas of the sun rising on the horizon, climbing to its zenith, and setting on the horizon at the point opposite its starting place. The relationship between daytime and light month is shown in figure 6.3.

*Section 3: The isomorphism of the alternation of the periods of
darkness and light with the alternate rotary motion of a primitive
drill or churn.*

The analogy between the sequence in the time periods shown in
figure 6.3 and the life of man is widespread, perhaps universal. Being
born corresponds to the beginning (sunrise, new moon), and dying
to the end (sunset, old moon). The period of light is the period of life,
and the period of darkness is that of death. As the period of darkness
occurring naturally in the lunar cycle has been transferred notionally
to the twelve dark days, the periods of darkness within the calendar
structure are (a) night, and (b) the twelve dark days. These twelve
days are devoted to the celebration of the chaos and reversal that have
been the subject of considerable study (Eliade 1955: 51-92; Babcock
1978). The oppositions that are concerned here – darkness, death,
and chaos set against light, life, and order – are familiar ones but it
is now possible to give them a more precise place in a calendrical
structure than it has been in the past.

The time before the commencement of the ordered seasons in the
Egyptian calendar was the period of the epagomenal days when the
gods Osiris, Horus, Seth, Isis, and Nephthys were born. Five days
were required to fit with the length of the solar year and so it is
possible to doubt that the birth story applied to this number of deities
in the archaic calendar. At any rate, for the present discussion of the
opposition between darkness and light, two of the gods are of special
importance, Seth and Horus, identified in other contexts as one of the
pairs of hostile twins who represent opposing principles often found
in mythology(Bianchi 1971; Leach and Fried 1975: 536, 1135). One
detail in the account of the birth of the gods given by Plutarch fits with
the 'twins' pattern; this is when he says of Typhon [Seth] that he 'was
born, not in the right time or place, but bursting through with a blow,
he leapt from his mother's side' (Griffiths, ed.,1970: 136-7). In the
listing of the epagomenal days, the birth of Horus actually occurs
before that of Seth, but in this comment on the birth of Seth it is
possible to trace a tradition of the reverse order of birth for it is clear
from parallels that the purpose of the violence on the part of the Seth
figure is to be born first of the pair, when, in the natural course of
events, the Horus figure would be born first.

In the Iranian instance of the myth, the hostile twins are Ohrmazd
and Ahriman who, in Zurvanite theology, are said to be the sons of
Zurvān (Time), who is referred to as masculine but is apparently

androgynous since 'he' gives birth to twins. The basic nature of the three figures as Time, Light, and Darkness is brought out by Zaehner: Since, then, it is clear that Zurvān cannot be the god of light, is there any ground for associating him exclusively with the darkness? The answer is equally plain: No. For this evasive deity is at once the god of light and of darkness; for he is the father of Ohrmazd and Ahriman who respectively dwell in the Endless Light and the Endless Darkness: or again he transcends the distinction between light and darkness, for these are only made manifest after the genesis of Ohrmazd and Ahriman whose attributes and places they are. So, in the folk-lore of Armenia which, as far as we can tell, still survives today, and which in this case is traceable back to ancient Irān, Žuk (Zurvān) or Žamanak (*zamān*, Time) is represented as sitting on a high mountain and as rolling alternately a white and a black ball of thread down the mountain-side. Here it is apparent that we have to do with a simple nature divinity, the god of the day and the night which mark the course of time. (1955: 56)

In Zurvanite, and also more generally in Zoroastrian ideology, Ohrmazd and Ahriman conduct their conflict over vast periods of time, but the above instance from folklore brings out their identity with the periods of day and night which it can be expected would be replicated in the periods of light and darkness in the year cycle. When the calendrical year is regarded as representing totality, it can imply all possible vastness within itself while still being a count of specific days, and so it is not reductive to bring the Iranian instance down to calendrical time, but it is perhaps because of the different time scale employed that the importance of the myth as a statement of calendar structure has been missed until now. The parallel with the Egyptian epagomenal days indicates that the birth myth is fixed at the time before the start of the seasons.

The myth itself tells how, in the beginning, Zurvān desired to have a son and made sacrifice to this end. After a thousand years, however, Zurvān had a moment's doubt about the efficacy of the sacrifice and Ahriman was conceived from this doubt while the desired son, Ohrmazd, was conceived from the sacrifice. Zurvān, knowing that two sons were in the womb, declared that he would make the elder of them king, and the inference is that he said this since he knew that, as one account states explicitly (Zaehner 1955: 66, 433), Ohrmazd lay between Ahriman and the exit from the womb and so Ohrmazd would in the natural course of events be the firstborn. However, when Ahriman got to know of Zurvān's vow, he broke from the womb through the navel, presented himself before Zurvān, and claimed the kingship. Zurvān was bound by his vow and made him king but limited the period of his rule, saying that at the end of the period Ohrmazd would reign (Zaehner 1955: 60-72,419-28; Watts 1969: 129-30).

Seth, it will be remembered, was born 'not in the right time or place' and broke from the womb forcibly, not using the birth passage. It seems from the Iranian parallel that the result of this violent action is that he is in the position of being able to seize the kingship but only for a limited period after which his brother will rule. It is easy to relate all this to the calendar structure already discussed. The first of the two brothers to be born, who is the god of chaos and darkness (Seth,Ahriman), rules over the period of darkness which lasts for the twelve days, the day of his birth evidently preceding this period and being the day particularly dedicated to darkness. Immediately after the period of darkness, the god of light (Horus, Ohrmazd) comes into power on the day particularly dedicated to light, and proceeds to rule the ordered period of the year with its twelve months divided into three seasons.

Seth and Horus, like Ahriman and Ohrmazd, both claimed sovereignty, and it appears that each of the two was particularly connected with rule over one of the 'two lands' of Upper Egypt and Lower Egypt (Griffiths 1960: 65-74). In the ceremony of 'Uniting the Two Lands' the gods are depicted holding either end of a cord knotted about a central column. The pose of the gods suggests that they are engaged in operating a drill by pulling first one way and then the other, as de Santillana and Dechend assert (1969: 162), and, although the knot is inappropriate to this activity, perhaps it can be understood as a token of reconciliation added to what was basically a drilling image. On the whole, there seems to be some likelihood that the representation indicates that the Egyptians had perceived the passage of time with its alternating periods of darkness and light under the rule of Seth and Horus as being analogous to the alternate rotary motion of a primitive drill or churn.

An implement of this kind indubitably occurs in the Indian myth in which the gods and demons (Devas and Asuras) churn the ocean by pulling alternately on either end of a serpent-cord wound about the world mountain. Brahmā is 'the impartial father of Devas and

Asuras alike, the twin parties, which, in the cosmos created by Brahmā, constitute the two rival groups', and the Indian myth equivalent to the birth of Ahriman and Ohrmazd from Zurvān is the creation by Brahmā of the 'alternate activity of pravṛitti and nivṛitti', that is to say, of 'turning forwards' and 'turning backwards' (Held 1935: 127-9, 138-47). Held comments:

[In the myth of the origin of death (Held 1935: 129; O'Flahert᠊ 1975: 37-43)] the existing world-order is represented as d pendent upon the uninterrupted sequence of life and deatl seen as an endless cycle of life fading into death and deat ᠊ constantly blossoming out again into new life, life and deat being called into existence simultaneously with the creation b᠊ the supreme Brahmāof pravṛitti and nivṛitti. We presume thaɩ the word, pravṛitti ('turning forwards') is suggestive of the turning of the churn-staff from left to right when the churning-cord is pulled by the gods, and the word, nivṛitti, as implying the 'turning back' again, the turning of the churn-staff from right to left when the churning-cord is pulled by the demons. (1935: 145)

The action of the drill or churn operated by the agents of the dual principles seems very likely to be a paradigm of time as represented in the calendar discussed here, and, if so, the succession of the years should be seen not as a series of complete cycles, as in true rotary motion where one circling action succeeds another with a constant spinning in the same direction, but as a series in which each year includes a double action where spin in one direction is succeeded by spin in the other.

Figure 6.4 shows this double action during the course of the year. The first day is the Darkness Day represented by the line at the end of the penannula to the observer's right. Then, reading counterclockwise, come the Twelve Dark Days which are shown in small figures outside the penannula opposite the equivalent months. After the twelve days comes the Light Day and, reading clockwise now, there occur, in order, the Spring Day, the two months of Spring, the Summer Day, the four months of Summer, the Year Day, the six months of Winter, and the Winter Day which brings the year to an end. This structure of the year is set out in figure 6.5 with a running count of days shown on the right.

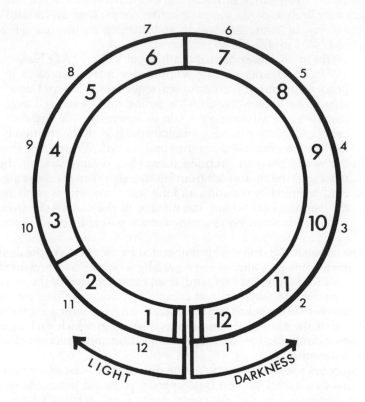

*Figure 6.4*

In conclusion, my hypothesis is that the period of the lunar year was covered by a subtle and tightly organised calendar in which such ideas as beginning, middle, and end, and short, intermediate, and long, as well as binary oppositions, were fixed in the structure itself. It may perhaps be surprising initially to find it suggested that the calendar followed the divisions of the human body but the idea of a spatial macrocosm mirroring the microcosm is familiar enough and,

| Darkness Day | 1 |
| Twelve Dark Days | 13 |
| Light Day | 14 |
| Spring Day | 15 |
| Spring (28 x 2) | 71 |
| Summer Day | 72 |
| Summer (28 x 4) | 184 |
| Year Day | 185 |
| Winter (28 x 6) | 353 |
| Winter Day | 354 |

*Figure 6.5*

on reflection, I think it will be realised that there is no reason why the portioning out of time should be excluded from the general mode of operation described by Cassirer:

> Wherever [myth] finds an organically articulated whole which it strives to understand by its methods of thought, it tends to see this whole in the image and organization of the human body. The objective world becomes intelligible to the mythical consciousness and divides into determinate spheres of existence only when it is thus analogically 'copied' in terms of the human body. (1955: 90)

It seems that the proposed calendar shape is fundamental, and that in a cosmological society it would have been an obvious form for the division of time to take. Of course, its various components would have been seen to correspond not just with the sections of the body and the social groups mentioned here, but with all the sets of items on the different levels of the paradigmatic structure of which it formed a part.

# 7. Distinctive Features in Cosmic Structure

A common objection to theories ... stems from the belief that they are limiting, blinkering and imprisoning devices. This belief confuses theory with dogma. A religious or moral dogma is something that it is proposed we live by – a scientific theory is something that it is proposed we live with and explore.

Bannister and Fransella 1971: 14

The theory that I invite others to live with and explore, as I have been doing myself, is that a single complex model underlies the major polytheistic religions of the archaic old world. The possibility that the model applies elsewhere should also be considered but I have, in this book, initially concentrated attention on the contiguous cultures extending geographically from Ireland to China. The term 'archaic' does not limit the model to a specific period of history or prehistory but does imply that, however recently evidences of the model may be found, it could function fully only in the context of what Toporov (1976) speaks of as the 'cosmological' kind of society which is atypical of the cultures of the western part of the land mass under discussion during the historical period.

The scholar who has done most in the past to bring out the structure of the polytheistic system is Georges Dumézil but, although I find his concept of the triad of the functions a valuable one which may well be relevant outside the Indo-European context where he studies it, I consider that he misunderstood the construction of the pantheon and that his theory at this point depends too heavily on the rather slender evidence of a single listing of Indic gods in the Hatti-Mitanni treaty documents of c. 1350 BC. I have two principal bases for my own theory of the organisation of the pantheon: macrocosmic-microcosmic structure, particularly as expressed in terms of calendar design (see chapters 5 and 6), and a theogonic sequence which articulates the deities in a family relationship (see chapter 11). What I am doing in the present chapter is looking at the composition of the proposed cosmic structure simply as structure, and showing that its complexity can be seen as resting on three simple oppositions or distinctive features.

Lévi-Strauss acknowledges the influence of Roman Jakobson's use of distinctive features in phonemics on his own work on binary oppositions in kinship systems and myth, and both scholars employed the same means in the face of 'a stunning multitude of variations' to uncover 'the invariants behind all this variety'.[1] I did not approach my subject, as they did, by way of theory, but certain oppositions emerged when I was exploring the particular cases of the calendar and the theogony. Now that they have been tentatively identified, it is very easy to express structure by way of them, and it is also easy to see how the concept of a complex pantheon could have been arrived at through the use of a basic thought process.

A total of eight distinguishable gods can be defined on the basis of three oppositions. The following general account of the process involved is taken from Jakobson (Jakobson, with Cherry and Halle 1971: 453), except that I give the numbers 1-8 where he has the letters A-H since I employ letters here for another purpose.

> A simple illustration of such a logical description is provided by [figure 7.1], which shows a set of eight 'objects' 1, 2, ... 8, to be identified by yes (+) or no (-) answers. Thus the group is first split in two, and we begin by asking, Is the object that we want on the right side (+) or not (-)? Successive subdivisions eventually identify any object in a set. If there are N objects in the set, and if N happens to be a power of 2, the number of yes-or-no answers necessary to identify each of the objects in the set is $\log_2 N$. The complete identification of any object is then a chain of plus and minus signs; thus, the object 7 in [figure 7.1] is identified by the chain (++-). ... The quantity $\log_2 N$ is conventionally expressed in 'bits'; the name for the unit is derived from 'binary digit' (i.e. yes or no choice).

The sequence in figure 7.1 is one of three bits, and this is all that is required in the study of the proposed pantheon.

For those unfamiliar with distinctive features, it may be helpful to consider a concrete example. If you are dishing out vanilla ice-cream, the choice is simply a yes-no one. You can either put vanilla ice-cream in a dish or not. The possibilities are exhausted with two dishes, one

|   | 1 | 2 | 3 | 4 | 5 | 6 | 7 | 8 |
|---|---|---|---|---|---|---|---|---|
|   | − | − | − | − | + | + | + | + |
|   | − | − | + | + | − | − | + | + |
|   | − | + | − | + | − | + | − | + |

*Figure 7.1*

containing vanilla ice-cream (+) and the other empty (-). When you have an additional choice of strawberry ice-cream, the possibilities are increased to four. If the possibilities are exhausted, one of the four dishes will contain both vanilla and strawberry (++), one will contain vanilla only (+-), one will contain strawberry only (-+) and one will be empty (--). With a further choice of peppermint, if the possibilities are exhausted, there will be eight dishes. One will be empty (---), while the others will contain vanilla, strawberry and peppermint (+++), vanilla and strawberry (++-), vanilla and peppermint (+-+), strawberry and peppermint (-++), vanilla only (+--), strawberry only (-+-), and peppermint only (--+). This can be expressed by listing the three features along with figure 7.1. Since the figure is a descent diagram intended to be read from top to bottom the sequence is quite clear, but I have added A, B and C so as to be able to use these letters to identify features in sequence in other contexts.

|   |   | 1 | 2 | 3 | 4 | 5 | 6 | 7 | 8 |
|---|---|---|---|---|---|---|---|---|---|
| A | vanilla | − | − | − | − | + | + | + | + |
| B | strawberry | − | − | + | + | − | − | + | + |
| C | peppermint | − | + | − | + | − | + | − | + |

*Figure 7.2*

I must at this stage, pending full discussion of the theogony, make the bald statement that, according to the present theory, the gods can also be defined in terms of distinctive features, although their case is rather more complex than that of ice-cream flavours. Since an entire universe is involved, each feature of the three in the sequence corresponds to a wide range of other qualities so that each god is or has a bundle of attributes which are far more than three in number. Nevertheless, there are basic distinctions within the cosmogonic sequence that seem to serve as central defining features. As I understand it, these are, in order, the oppositions high/low, dry/wet, and light/dark, and the sequence can be shown as in figure 7.3.

| | | 1 | 2 | 3 | 4 | 5 | 6 | 7 | 8 |
|---|---|---|---|---|---|---|---|---|---|
| A | high | − | − | − | − | + | + | + | + |
| B | dry | − | − | + | + | − | − | + | + |
| C | light | − | + | − | + | − | + | − | + |

*Figure 7.3*

It will be seen that high, dry and light are treated as the positive terms. This point is quite clear in the cases of high and light, but the opposition of dry and wet appears to be a reciprocal rather than a hierarchical one and the decision to take dry as the positive term rests on correspondence and will require to be discussed elsewhere. One proposed set of correspondences should be noted here, however, so that the present statement can be related to my calendar studies. When the lunar phases are regarded as the prototypical form for the divisions of time, the temporal series taken to correspond to the one above can be expressed as: A waxing moon, B maximum moonlight (i.e. the period including the full moon but not the crescents), C moonlight (i.e. the period of the moon's visibility as opposed to the dark of the moon). Such a correspondence of two sequences – the cosmogonic sequence and the calendar sequence – would allow discoveries in one area to provide help in deciphering the code of the other. A correspondence system has to be approached with caution since inappropriate analogies may leap to the eye and have to be set aside but it is not without in-built constraints and it is through the constraints that we are enabled to receive meaning – language rather than babble. Major tasks in the exploration that lies ahead are the definition of the operation of constraints and the recognition and rejection of the false steps in the handling of an intricate network of correspondences that can be as glaring and harmful as illogical steps in an argument.

That remark can serve to introduce Aristotle who is more of a Janus figure than is sometimes thought. His contribution to logic is widely recognised, but from the point of view of the study of cosmology it is his unargued statements that are of particular value for they can give an insight into the cosmological basis of his thought. He holds, for example, that 'above is more honourable than below' (Lloyd 1966: 52), and his theory of the four elements relates to this presupposition. The composition of the elements of water, earth, air, and fire rests on the two opposites ±hot and ±dry. As the two hot elements, fire and air, move upwards, and the two cold elements, earth and water, move downwards, hot can be seen to correspond to above or high and is the positive term. The situation is less clear in the case of the dry/wet pair but G.E.R. Lloyd's study of the point (45-6, 61-2, 64) indicates that Aristotle treats dry as the positive term in the context of series of paired opposites. The opposites can accordingly be laid out (figure 7.4) as in the previous figures, the composition of the various elements being water (--), earth (-+), air (+-) and fire (++).

| A | high (hot) | − | − | + | + |
|---|------------|---|---|---|---|
| B | dry        | − | + | − | + |

*Figure 7.4*

The Hippocratic writers treat the four humours as being composed of the same opposites: blood is hot and wet, yellow bile hot and dry, black bile cold and dry, and phlegm cold and wet. This is a microcosmic concept dealing with the human body, but, as the Hippocratic writers are concerned with the correct medical treatment at each of the various seasons of the year, they also deal with the temporal macrocosm and note that spring is hot and wet (+-), summer hot and dry (++), autumn cold and dry (-+), and winter cold and wet (--) (Schöner 1964), so giving a cycle which can be represented as in figure 7.5 with the polarities lying along different axes.

These Greek writers are talking in quasi-scientific terms and do not include the gods in their structures, but it is clear that their oppositions apply to the entire cosmos. The sets of the elements and of the humours and seasons illustrate the case of cosmic structure

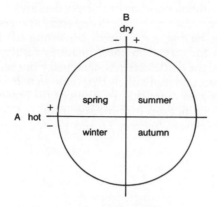

*Figure 7.5*

divorced from gods. It will also be necessary to study the case of the opposite extreme – gods or their euhemerised substitutes divorced from cosmic structure – as well as the intermediate case of a structured ensemble of gods in a cosmic setting.

As regards terminology, it will be useful to speak of 'the three axes of polarity' (A ±high, B ±dry, C ±light). Two axes can be shown as in figure 7.5. It is not possible to show a third axis on a flat plane, but a method of presenting the third opposition (C) on the same figure is to divide the quarters in half and to shade the negative (or dark) half of each, as in figure 7.6.

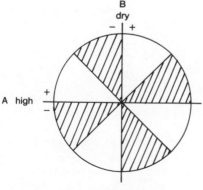

*Figure 7.6*

The proposed pantheon consists of sharply demarcated deities of a kind that it has been customary to call departmental gods. I prefer to think of them as having (or, indeed, consisting of) bundles of attributes; this lays the stress on their qualities rather than their activities. Of course, the deities are not detached from society and it has come to seem more and more likely that the system has at its focus a pair of ruling gods who share the attributes of the others and that, whatever other social functions it may have, it serves to validate the institution of kingship (see chapters 11 and 12).

# 8. The Design of the Celtic Year

I will begin with some remarks on the design of the Icelandic calendar about which Professor Hermann Pálsson spoke to the Traditional Cosmology Society in May 1984. Both the Celtic and Icelandic calendars have notably clear and simple outlines which make them excellent subjects for the study of design. I am not concerned here with intercalary periods and will exclude them from consideration.

It will be useful in the discussion of design to define a year according to the points of division between one period and another. It was clear from Professor Pálsson's discussion that the Icelandic year is, and has been, apprehended as divided shortly after the equinoxes into halves – summer and winter – and one would accordingly think of it in the first place as a two-point year. There are indications, however, both in the present calendar structure and in the historical record that the design is rather more complex.

Professor Pálsson remarked in passing that, according to Tacitus, the Germans had a three-season year consisting of winter, spring, and summer (*Germania* 26, trans. Hutton 1914: 300-1): 'Winter, spring, summer have a meaning and name; of autumn the name alike and bounties are unknown.' He could not relate this assertion to the Icelandic calendar but I suggest that it is feasible to trace out a connection. The Icelandic calendar was basically derived from Scandinavia and a reference in the *Ynglinga Saga* which is appropriate to a three-point year may serve as a link between the statement of Tacitus and the structure of the Icelandic calendar (Sturluson, *Heimskringla*, trans. Laing, rev. Foote 1964: 13): 'On winter day [mid-October] there should be blood-sacrifice for a good year, and in the middle of winter for a good crop; and the third sacrifice should be on summer day [mid-April], for victory in battle.' Two of these points coincide with the points of division between summer and winter already noted. The third is marked in Icelandic custom as an extended festival period in December-January running from the month Ylir which ends before Christmas to 1 Þorri which is *miður vetur* (the middle of winter), half-way between winter day and summer day (Bjornsson 1980: 69, 73-4, 78-80). Since this period, which will be referred to here for convenience by the name 'Yule', is a continuous festival, it forms a single point of division. The Icelandic year, then, can be seen to fall into three parts: the first half of winter preceding Yule, the second half of winter following Yule, and summer, and I suggest that it is these three parts that correspond to the sections called by Tacitus winter, spring, and summer. When I say 'correspond' this is not to say that the actual periods of the solar year were the same but that the three periods take the same place in the design.

It is the concept of design that opens up new possibilities in the exploration of calendar structures. The connection of the different elements of the design with the year is variable and, looking at European calendars alone, has varied so much that there have been no recent attempts to relate different systems to each other. The analogy of house design may be a helpful one in clarifying relationships, although I am not suggesting that this is a traditional analogy like the ones I will speak of later. In terms of this analogy, the builders are permitted to construct houses with rooms of different sizes but have to meet the specifications laid down, which are that each house must contain a kitchen, a bathroom, and three other rooms, one of which can be designated as a bedroom and the other two as public rooms. It is possible to think of the year structure numerically in different ways but in the present terms there are five components, three of which are the seasons, 1 spring and 2 summer (the public rooms) and 3 winter (the bedroom). The Yule period in the Icelandic calendar is the bathroom. The kitchen, which is the representation of the female within the year, may not show up in a calendar structure as it can consist of a single day, but I have hypothesised the presence of this feature at the point of division between summer and winter (chapter 2) and there is no difficulty in identifying this spot in the Icelandic calendar although its possible connection with the female in this context has yet to be investigated. Four of the five proposed components in the design, then, are evident in the Icelandic calendar and the remaining one may be latent there. In figure 8.1, the three 'season' sections are numbered, the Yule period is shaded, and the postulated place of the female is marked F.

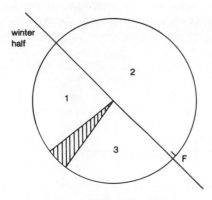

*Figure 8.1*

I have suggested in chapters 5 and 6 that the year cycle in the Indo-European context and beyond is treated as isomorphic with the lunar cycle, and I find the Roman treatment of the lunar cycle particularly helpful. The Roman month is a three-point one, the divisions coming at the Kalends (equivalent to new moon) on the 1st, the Nones on the 5th or 7th, and the Ides (equivalent to full moon) on the 13th or 15th. I suggest that these three divisions correspond to the three seasonal divisions of the year, and that the extended festival period that precedes spring corresponds to the dark of the moon, the interlunium. The point between the period of the month equivalent to summer and the period of the month equivalent to winter is at full moon which I understand to be distinguished from the rest of the month as the place of the female. The Roman year calendar is so packed with festivals that it is difficult to isolate major divisions, but one that it is possible to distinguish and which has some importance in the discussion of the Celtic year is the year-beginning at the beginning of March which preceded the shift to the familiar year-beginning in January (Michels 1967: 18, 97-101, 121-30). Figure 8.2 shows the three Roman month sections and the features of the interlunium and the full moon. It also indicates the year-opening on 1 March at an equivalent point to new moon, and shows the opposite point of the year as 1 September. The splitting of the waxing half is unequal and in this case the division point in the year cycle is given as 1 May which may either be a precise equivalent, as one approach to the lunar day-count suggests (see chapter 6), or a rough approximation.

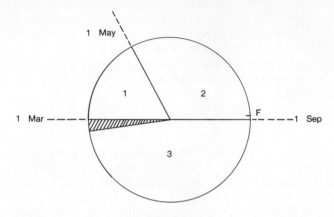

*Figure 8.2*

I have spoken of the five components as forming parts of a design as kitchen, bathroom, bedroom, and two public rooms could be the specified parts of a house. I am not suggesting that this analogy is any more than a casual, modern one which helps us to cope with the unexpected idea that seasons may vary in size, and, although the year structure would have been related to buildings, it would not have been in just this way. There are other analogies, though, that I wish to offer as ones that were closely tied to the divisions of the year. One that I have already explored is the analogy of the human body with three parts corresponding to the three seasons: head to spring, upper body to summer, and body below the waist to winter. I want here to consider the correspondence with the human life cycle. We can speak of the new year and the old year but it is not really the year that is new or old, as we can see clearly from the fact that different points in the year can serve as the period of end-and-beginning. Historically, the year has proved malleable and has taken its impress from that to which it has been analogically tied – in part, as already indicated, the lunar cycle, in part the microcosmic body divisions, and in part the human life cycle. The new year, equivalent to new moon, corresponds to human birth, and the old year, equivalent to old moon, corresponds to the moment of death. The extended festival period between old and new year equivalent to the interlunium corresponds to the state of being dead.

At this stage in the discussion of the projection of the human life cycle on to the year cycle, it seems necessary to distinguish the male from the female. It is when we see the new year as equivalent to male birth and old year to male death that we appreciate the necessity of the placing of the female at a half-way point when the male makes the transition from the single to the married state. The further division point between birth and marriage can be understood as corresponding to the other rite of passage at puberty when the boy joins the ranks of unmarried men. The five components, then, would be boyhood, manhood before marriage, marriage, manhood after marriage, and the state of death. In my full study of calendar in chapter 6, all points of transition within the year are equated with special festival days, but the festival day of marriage is different from the others since it involves the female. The distinct place of the female in the system is called into play in the Celtic calendar to which we can now turn, merely noting first that, if it is the human life cycle that is being treated, as I suggest, it is fairly certain that, while components may be manipulated at will to fit other requirements of the year cycle, it is improbable that any will be left out entirely.

A requirement that appears to have been felt in the case of the Celtic calendar was a relationship with the main solar points of the year, the solstices and the equinoxes. The Celtic year is a four-point one and, while the four festivals of Samain (1 November), Imbolc (1 February), Beltaine (1 May), and Lugnasad (1 August) do not fall on the main solar points, they do bracket them,[1] so that the solar points come in the middle of seasons of which the four festival days mark the beginnings.

Another feature is that each of the year quarters is equivalent to a world quarter, i.e. the spatiotemporal correspondence to be expected of a cosmological system is a very straightforward one in this instance. The division of space is especially clear in the case of Ireland, which, as an island, lent itself readily to the establishment of lasting spatial categories. Four provinces lie in the four directions: Leinster in the east, Munster in the south, Connacht in the west, and Ulster in the north. The provinces fall into two divisions just as the four seasons fall into summer and winter halves with division points at Samain and Beltaine. The superior half of Ireland, the half of Conn, consists of the provinces of Connacht and Ulster[2] and a passage in *The Death of King Dermot* implies the correspondence of this half with summer. Dermot, i.e. Diarmait mac Cerbaill who reigned c. 544-565 AD (Byrne 1973: 87-105), is described as holding the feast of Tara at Samain, the beginning of the Celtic year, and then setting out on a historically anachronistic but cosmologically informative circuit:

> Then on his regal circuit Dermot [set out and] travelled right-handed [i.e. south and west about] round Ireland, that is to say: from Tara into Leinster; thence into Munster; thence into Connacht, and athwart Ulster's province; so that at the end of a year's progress he would by *samhain* again reach Tara in time to perform his *samhain*-tide office and to meet the men of Ireland at Tara's festival.[3]

The implication is of a steady progress through the four quarters of Ireland made in such a way that movement in space corresponds to temporal progress through the four seasons as shown in figure 8.3. Winter corresponds to the inferior half (-A) and summer to the superior half (+A).[4]

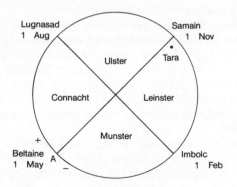

*Figure 8.3*

The division into four quarters goes back as far as there is direct calendar evidence, but there are indications of a former three-point year. As J. Loth noted, for example, 'There are only three out of four names that designate [the Celtic divisions of the year] that are genuinely Indo-European: those that designate winter, spring and summer'.[5] The following discussion is based on the hypothesis that the year had been on the same model as the Roman three-point scheme illustrated in figure 8.2. A transition from a year divided in that way to the one found in the Icelandic calendar would have been a simple one involving merely a variation in the length of the seasons since the interlunary period is represented by Yule. The Celtic calendar differs here for, although there are traces of a special twelve-day period equivalent to the interlunium (Rees and Rees 1961: 93), these seem to be only echoes of a former state. The transition from the three-point scheme shown in figure 8.2 would, in the Celtic case, have had to be a complex one, and I lay out the suggested adaptation below in three stages for the sake of clarity although I regard the stages merely as aspects of a single comprehensive restructuring.

*Stage 1.* Given that the female point of the year is in correspondence with one of the world quarters as I have argued in chapters 1 and 2, it would have been easy to create a four-quarter year by giving a quarter to each of the three seasons and equating the female with the other quarter. Only one point of the four-point year coincides with a point of the postulated three-point year – 1 May. A quartering taken from that point inevitably places the other four season-openings at 1 August, 1 November, and 1 February, in terms of our present calendar. This stage eliminates all the asymmetrical features of the calendar apart from the period equivalent to the interlunium which is indicated by shading in figure 8.4, which shows the stage 1 change.

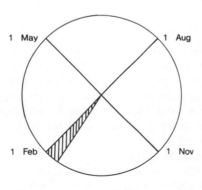

*Figure 8.4*

*Stage 2.* This stage, I suggest, eliminated the asymmetrical feature of the set of days equivalent to the interlunium by notionally expanding this 'dark' period so that it coincided with the entire half-year from November to May.[6] A change of this kind would explain the occurrence of the year-beginning at Samain. The Indo-European custom was to reckon darkness as preceding light and the year would have begun at the start of the short period of dark days and have been followed at no long interval by the beginning of the period of light. What we appear to have in the case of the Celtic year is a year-beginning at the start of an extended dark period at 1 November and the beginning of the light period six months later on 1 May, as shown in figure 8.5.

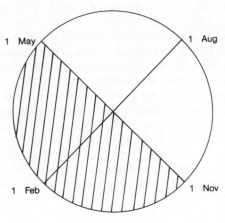

*Figure 8.5*

*Stage 3.* I suggest that a further change was made, motivated by avoidance of correspondence between any part of the newly created dark half of the year with the superior half, +A, and that the dark half of the temporal scheme was brought into correspondence with the inferior half of the spatial scheme. Keeping the spatial model static and showing the other revolving to match it, the change may be shown as in figure 8.6.

These suggested changes result in a symmetrical system with the dark half of the year corresponding to the inferior half, as found in *The Death of King Dermot.* It should be noted, however, that, since the bright half of the year, which had formerly consisted of spring and

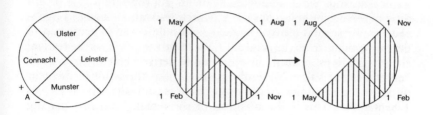

*Figure 8.6*

summer (equivalent to the half of the lunar cycle beginning at new moon), had, as a result of the change, come to consist of summer and autumn, the proposed shift is a radical one. It may, I think, have resulted in a series of hybrid ritual or customary observances, with the transfer of associations from the three-point to the four-point system taking place along the lines indicated in figure 8.7.

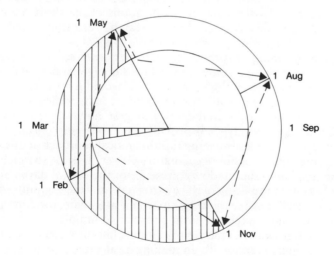

*Figure 8.7*

The adjustments can be outlined in this way. 1 May (Beltaine) could be expected to retain elements appropriate to that time in the natural seasons while acquiring elements appropriate to the beginning of the bright half from the 1 March festival. 1 August (Lugnasad), as the second festival in the bright half, could be expected to derive elements from the original 1 May festival, while, as the festival closest to harvest, it could be expected to derive elements from the 1 September festival. 1 November (Samain), as the point in the year opposite the beginning of the bright half, is equivalent to the earlier 1 September. This is the position of the female, and this placing accords with the celebration of the king's nuptial feast at Samain. 1 November is, however, also involved in the associations of the opening of the period of nominal darkness that originally preceded 1 March which have now, with the increase of the length of time devoted to darkness, to be carried across to a distant point of the year. Samain therefore combines powerful elements – marriage and death – which were formerly celebrated six months apart. 1 February (Imbolc) is the leanest of the festivals, and it is possible to speak, as D.A. Binchy does (1958: 113), of Beltaine, Lugnasad, and Samain as 'the three main festivals of the pagan year'. This is explicable if the associations of 1 March as the beginning of the light half were transferred to 1 May, and the associations of the short period of darkness immediately preceding 1 March were transferred to 1 November, leaving only a fraction of the original associations of the 1 March beginning-of-spring festival to be transferred to 1 February.

Since the proposed adaptation is based on a hypothetical form of an earlier calendar, it might seem that it would have been wiser to defer this rather intricate development of a hypothesis until the hypothesis could be put to the test. I suggest, however, that study of the Celtic festivals may itself provide a suitable testing ground, and that it is worth while absorbing this theoretical approach at this stage and trying out its possible explanatory power. There is so much that it does explain – the shadowy presence of the twelve days and the three seasons, the placing of the quarter days and the beginning of the year, the celebration of the king's marriage at Samain, the dominance of Samain, and the relative unimportance of Imbolc.

The design of the Celtic year is unique, but the Celts were not, of course, acting exceptionally in making a calendar change, and they can probably be regarded as taking part in a general movement away from the asymmmetrical towards the symmetrical. There need have been no change in religion involved, and the way I have worked with calendar here has assumed a continuing tradition which altered the outward shape but carefully incorporated elements of the former meaning in the new design. The question of the date of the change needs further study but it can be said that the four-season year was firmly entrenched in both the British and Goedelic branches of the Celts in the British Isles and that the Gaulish Coligny calendar of the first century AD is in keeping with a fourfold rather than a threefold scheme. The difference between the lunar year with intercalation found in the Coligny calendar and the solar year found in later tradition is not a difference in design as the term is used here. Modification from a quartered lunar year to a quartered solar year merely involves a slight addition to the length of each season. The suggested major change, the one which resulted in the use by the Celts of a four-season year, apparently took place before the time of the Coligny calendar.

# 9. Polarity, Deixis, and Cosmological Space and Time

Earlier (see especially chapter 2), I have posited the correlations shown in figure 9.1 between the divisions of the upright human body, the cosmic levels, and the structuring of an archaic calendar.

| | | |
|---|---|---|
| head | heaven | spring |
| upper body | atmosphere | summer |
| WAIST | EARTH | HARVEST |
| lower body | netherworld | winter |

*Figure 9.1*

These correlations were suggested by Indo-European sources, but I do not claim that the correspondences were necessarily confined to speakers of Indo-European languages. In fact, leaving aside the specific feature of the subdivision of one of the halves and concentrating attention on the two halves and the mid-point, recent work in linguistics and psycholinguistics indicates that the structuring has a basis in human perception and so the suggestion that comparable structuring may be widely found throughout the world can be offered.

Scholars who have been particularly responsible for the advances of thought drawn on here are Herbert and Eve Clark and Elizabeth Closs Traugott at Stanford University and John Lyons, Professor of Linguistics at Sussex and formerly at Edinburgh.[1] One of the illuminating key concepts they handle is polarity and another is deixis. I will consider them here in turn.

Right-left has played a prominent part in anthropological discussions of duality (see Needham 1973), but in the psycholinguistic studies of perception (see H. Clark 1973: 35; H. and E. Clark 1977: 535; E. Clark 1979: 53), it is beaten into third place by up-down and front-back. Linguistically, Lyons points out, the up-down dimension is primary and both it and the front-back dimension are more salient than the right-left dimension. He continues (1977: 2.691):

> In the up-down and front-back dimensions there is not only directionality, but polarity: what is above the ground and in front of us is, characteristically, visible to us and available for interaction; what is beneath the ground or behind us is not. Upwards and frontwards are positive, whereas downwards and backwards are negative, in an egocentric perceptual and interactional space based on the notions of visibility and confrontation.

Similarly, Herbert Clark states in his discussion of perceptual space (1973: 35): '(1) ground level is a reference plane and upward is positive; (2) the vertical left-to-right plane through the body is another reference plane and forward from the body is positive.' It is essential in the study of cosmology to note that, while the second reference plane is expressed in terms of the microcosm of the human body, the first is macrocosmic; as Lyons puts it (2.690), we have 'a fixed zero-point at ground-level'. I have not seen the corresponding zero-point in the microcosm identified in published work by scholars working on perception, but in an Edinburgh Ph.D. thesis in linguistics, Marilyn Jessen notes that the zero-point in terms of the human body is at waist level and that, again, upwards from that point is positive and downwards negative (Jessen 1975: 82-3). This finding can be paralleled in ethnography, (see e.g. Littlejohn 1973: 289-90 and Duff-Cooper 1985: 133-5.) There are properties of the body that lend themselves to this polarisation, but the polarised body is probably also to be understood as an analogue of the macrocosm where the perceptual differences establish the polarity.

This approach through psycholinguistics clarifies the spatial correspondences shown in figure 9.1. Waist and earth's surface are zero-points without vertical extent, while upwards in the positive direction lie the zones of the upper body and head, the atmosphere and heaven, and below in the negative direction lie the zones of the lower body and the netherworld. I have also suggested (chapter 2) that the zero-point of the polarity was distinguished from the positive and negative zones by being the 'here' in contrast to the 'elsewhere' and by being associated with the female in contrast to being associated with the male. Psycholinguistic studies have no bearing on the second suggestion but are most illuminating with respect to the first, and it is now relevant to turn to consideration of deixis.

This term is in general currency in linguistics but, as it may be strange to those unfamiliar with this subject, it may be helpful to begin with a definition by Miller and Johnson-Laird which runs as follows (1976: 395): '"Deixis" – which is the Greek word for pointing or locating – refers to those aspects of a communication whose interpretation depends on knowledge of the context in which the communication occurs.' Words such as 'here', 'there', 'this', 'that', 'now' and 'then' are deictic. We don't know, for example, where 'here' is without knowing the location of the speaker. Ego is at the centre and is the landmark to which reference is made. To interpret deictic language, we need to know both the location and the direction of facing of ego, and also, as Lyons points out in the following passage, we need a temporal fix (2.638):

> The canonical situation-of-utterance is egocentric in the sense that the speaker, by virtue of being the speaker, casts himself in the role of ego and relates everything to his viewpoint. He is at the zero-point of the spatiotemporal co-ordinates of what we will refer to as the deictic context. Egocentricity is temporal as well as spatial, since the role of speaker is being transferred from one participant to the other as the conversation proceeds, and the participants may move around as they are conversing: the spatiotemporal zero-point (the here-and-now) is determined by the place of the speaker at the moment of utterance.

The zero-point in space is 'here' and the zero-point in time is 'now' and the anchor is the speaker, ego.

However, it is possible to select another point than ego as the anchor both linguistically (Lyons 2.677; Traugott and Pratt 1980: 276) and otherwise; that is, it is possible to have what I shall call a transferred deictic zero-point. Norberg-Schulz, in an architectural study, deals with the spatial transfer to an externalised world centre (1971: 18):

> In terms of spontaneous perception, man's space is 'subjectively centred'. The development of schemata, however, does not only mean that the notion of centre is established as a means of general organization, but that certain centres are 'external-

ised' as points of reference in the environment. This need is so strong that man since remote times has thought of the whole world as being centralized. In many legends the 'center of the world' is concretized as a tree or a pillar symbolizing a vertical *axis mundi*.

The subjective centring of space may perhaps imply that such a perceptual polarity as above and below relative to the earth's surface is realised in egocentric terms, as if a human figure – a transferred ego – were in position at the zero-point. I take it here that this is a correct interpretation, but will discuss the matter of deictic centring more fully in the context of time.

As shown in figure 9.1, I hypothesised a year with harvest as a mid-point preceded by spring and summer as one half and succeeded by winter as the other half. The correspondences suggest that this structure is to be understood in the same way as the spatial ones with harvest as zero-point, the summer half as positive and the winter half as negative. However, the year has come to seem to me an especially receptive temporal cycle which historically has received the impress of different shapings of symbolic meaning even among peoples speaking Indo-European languages (see chapter 8), and the zero-point appears to be fixed in the year cycle at least in part through the operation of analogy and to be firmly located only in the smaller diurnal and monthly cycles. It will be better, therefore, to concentrate attention on them and see what an approach through language has to offer in relation to these time periods.

Linguistically, spatial expressions are more basic than temporal expressions and there is an 'interdependence of time and distance (in that what is further away takes longer to reach)' (Lyons 2.718). Movement in space can be observed very clearly in relation to the passage of time in the daily course of the sun. Deictically, an object in motion is conceived of as either 'coming to' or 'going from' ego or a transferred deictic zero-point and in a statement dealing with both space and time Traugott points out that polarity is involved:

> *Come* involves motion toward a goal where the speaker is, *go* motion toward a goal where the speaker is not; that is, *come* is positive, *go* negative (cf. Fillmore, 1972). ... In the temporal system, *now* is the normal, semantically unmarked state. Motion toward it is positive, motion away from it negative. (Traugott 1975: 216; and cf. Givón 1973: 918)

The sun's course can be conceived as a single journey from the eastern horizon (A) as source to the western horizon (B) as goal, and in this conception the noon position is merely one point like any other on the path, but it can also be conceived deictically with the noon position as deictic zero-point (X) as a journey from the eastern horizon as source to the noon position as goal and from the noon position as source to the western horizon as goal, as shown in figure 9.2.

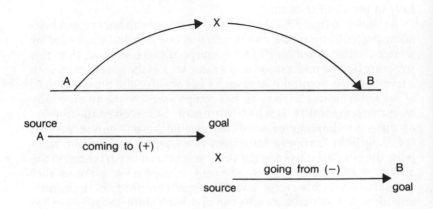

*Figure 9.2*

If ego is regarded as projected into the noon position, the sun will be viewed as coming in the morning and going in the afternoon, and the morning will be positive and the afternoon negative.

There is a clear mid-point in the sun's daily course to which it ascends in the morning and from which it descends in the afternoon and there is also a clear mid-point in the lunar cycle – the full moon. It is probably relevant that over the period of its visibility the moon is seen to undergo changes that are analogous to changes observed visually in objects coming and going. Taking the zero-point as the night of full moon, on each night that passes as an approach is made to this point the moon is bigger and corresponds to an object which looms larger as it approaches ego and similarly the moon dwindles each night after full moon just as an object appears to diminish in size as it goes from ego. Polarities of this kind in the course of the day and the month may be illustrated by two specific instances. In Chinese tradition, in the pair morning-afternoon morning was yang (light, +) and afternoon yin (dark, -) (de Groot 1901: 53), and, in Indian tradition, 'the fortnight beginning with the new moon was called the bright half (*śuklapakṣa*) and the other the dark half (*kṛṣṇapakṣa*)' (Basham 1967: 494). Deixis offers a possible explanation of these contrasts.

The subject of deictic polarities in cosmology calls for much more investigation both in earlier literature and in the field, but even on the basis of the present brief study it is possible to make the following statement which I find relevant to the posited archaic old world structure with which I am working, and which may well have a broader application.

The deictic zero-point in space is at ground level for the macro-cosm and at waist level for the human body, and the space above the zero-point is positive and the space below it negative. The deictic zero-point in time is at noon for the diurnal cycle and at full moon for the lunar cycle with the time before the zero-point positive and the time after it negative. The zero-point may be applied analogically to various times in the year but, whenever it occurs, the period before it in the annual cycle will be positive and the period after it negative.

It should be added that, although the 'zero-point' is appropriately named so far as polarity is concerned, it is far from empty, and in fact serves as the pivot of the system.

# 10. Cyclical Time as Two Types of Journey and Some Implications for Axes of Polarity, Contexts, and Levels

*The journey as metaphor for cyclical time*

As Leopold Howe comments in a recent article (1981: 22): 'a study of the way that a people perceives time can only be accomplished by an investigation into the ways in which the passage of time is reckoned, how the intervals are obtained, the systems by which such units are counted, if in fact they are, how the units are conceptualised and what images and metaphors are employed'. Recent work on metaphor has made us very conscious that those of us who are engaged in studying are also enabled to perceive and express different things depending on the analogical model employed. In order to avoid being trapped by a model, it seems advisable to explore any latent model explicitly and to move flexibly from one model to another to glean information on different aspects of the subject under investigation. I shall use several metaphors here in exploring cyclical time but shall take as my base the idea of the journey.

Journeys take place in space, and time and place seem to be closely interlinked in human experience. Geographers have recently been paying increasing attention to time and the geographer Yi-Fu Tuan has an interesting formulation of cyclical or, as he calls it, 'astronomic' time (1978: 7, 14):

> Astronomic time is experienced as the sun's daily round and the parade of the seasons; its nature is repetition. Mythic space is often organized around a coordinate system of cardinal points and around a central vertical axis. ... Has the human mind perceived and organized space in accordance with the biological experience of rhythmic time? Rhythm or period has two models: the pendulum that swings back and forth along a line, and the clock whose hands move in circles.

A cycle is a recurrent pattern at the end of which one is returned to the starting point from which the next instance of the recurrent pattern will begin (cf. Howe 1981: 231; Bloch 1979: 166). The two models suggested by Tuan allow an important distinction to be made between two concepts of cyclical time: in the case of the circle described by the hand of the clock the return to the starting point is made without traversing the same ground again whereas in the case of the oscillation arc described by the pendulum the starting point is reached only when a return has been made along the same track. Leach, in studies of time in *Rethinking Anthropology* (1961), rather confused the issue by using the pendulum image in loose association with the concept of alternation. Alternation is compatible both with the clock circle model and the pendulum arc model, as will be apparent in the discussion below. What is present in both the models selected by Tuan that is not present in the simple idea of alternation is the moving element – the clock hand or pendulum that represents the constantly shifting 'now' of experienced time. At any time, one could indicate 'when' one is in the system, or 'where' one is in the spatial analogue that can be presented diagramatically. The 'now' runs through the system and I find a rather apt metaphor for it in the world of word-processing; the 'now' is the cursor, the lit spot that indicates where one is operating at any given time. But 'cursor' is the Latin for 'runner' and returns one to the more immediately human image of a person at the 'now' point running a course or, if the speed and means of locomotion are not specified, making a journey. That journey out and back to the starting point of the cycle can be of two kinds. The traveller can either cover fresh ground throughout (a circling journey) or can return along his own tracks (a reversing journey). I shall now look at instances of these two different concepts of cyclical time as held by specific peoples – the Vedic Indians, and the Rindi of Indonesia.

*The reversing journey (Vedic Indians)*

On the evidence of the *Rigveda* and other Indian texts one can say that the total period of day-and-night was represented by the metaphor of a journey by a traveller who returns along his own tracks, the traveller being the sun. Gregory Nagy explores some aspects of this idea in a recent article (Nagy 1980: 167):

> As we begin to examine the traditions about this specialized sun-god Savitṛ, we note that there are contexts of darkness as well as brightness. Savitṛ protects the righteous *at night* (*RV* 4.53.1) and wards off the demonic Rakṣas-es *all night* (1.35.10).

The time of these Rakṣas-es is the night (*RV* 7. 104.18); in the East they have no power, because they are wiped out by the rising sun (*Taittirīya-Saṃhitā* 2.6.6.3). As the *Rig-Veda* makes clear, the daily breakthrough to light in the East is caused by Savitṛ (10.139.1). How, then, can the sun-god Savitṛ be present at night? As we see from *RV* 1.35, the Savitṛ hymn, the god reverses the course of his chariot with each sunrise and sunset: he travels through the brightness at day and then through the darkness at night. During the night-trip, he is the good aspect of darkness, just as he is the good aspect of brightness during the day-trip. The night-trip of Savitṛ is especially precarious for mortals not only because of the darkness but also because the daylight course of the sun is reversed. After the forward course of the chariot at daytime comes the backward course at nighttime. This forward/backward movement of Savitṛ is expressed in terms of downstream/upstream, the words for which are *pravát-/udvát-* (*RV* 1.35.3):

> *yāti deváḥ pravátā yāty udvátā.*
> 'the Deva goes downstream, goes upstream.'

The sun's chariot was envisaged as having a bright wheel on the right side and a dark wheel on the left side so that it was visible on earth only when the chariot made its day journey from the east to high in the south to the west. This is the movement to the right in the direction of the sun. When the return journey was made in the left-handed direction from the west to high in the south to the east the dark wheel was towards the earth and the sun was invisible. The idea was also expressed without the image of the chariot. The sun was conceived of as a disk with one bright face and one dark face. During the day the bright side was towards the earth and at night, when the sun was on its reverse course, the dark side was towards the earth.[1]

In his discussion of this concept, J.S. Speyer remarked that 'it may be observed that the supposed returning course of the sun at night, from the west to the east through the south ... agrees very well with the religious practices always followed in the ritual pertaining to the *piteras*, to Rudra, in the *abhicāra*-rites, and in all other performances which have in view the beings and spirits of night and darkness' (Speyer 1906: 727). Speyer has in mind the ritual movement to the left associated with darkness and the dead, and his comment suggests that the concept of the reversing journey of the sun was not an isolated fancy but was integral to a system. Indian tradition also gives us the myth of the churning of the ocean where two teams of gods, the Devas (including Indra) and the Asuras (including Yama, god of death), pull alternately at either end of a snake wound about the world mountain in an action where the two circling movements of turning forwards to the right (*pravritti*) and turning backwards to the left (*nivritti*) succeed each other. I have used this concept in chapter 6 where I have worked out some possible implications of the reversing journey for the year calendar. One point to note is that darkness is reckoned to come before light in the Indo-European tradition and so the movement to the left should be understood to precede the movement to the right.

In chapter 9 I looked, in the Indo-European context, at the passage of the sun across the sky and identified a deictic zero-point (a fixed 'here and now') at noon which corresponded to the full moon in the lunar cycle. This pivotal point related to the female. When one sees the whole sequence of day and night, this pivotal point (marked X below) is approached twice in the course of a reversing journey, first from one direction and then from the other. Figure 10.1a shows the double passage of the sun across the sky beginning with the movement to the left which is that of the night journey of the sun. Figure 10.1b shows the same movement in relation to the vertical structure also discussed in my deixis chapter. Here the pivotal point comes at the level of the earth's surface and the waist. Heaven and the head are at the source point associated with the beginning of the light period and so the total movement starting with night begins from the feet, with the movement from bottom to top corresponding to movement to the left, and the movement from top to bottom corresponding to movement to the right. Figure 10.1c shows a circling movement to the left preceding a circling movement to the right. This is the representation I used in my study of the year calendar (chapter 6).

Let us look at the main polarities here before moving on to the comparative instance of the Rindi. Movement to the left is negative, movement to the right positive; night is negative, day positive; movement up (or upstream) is negative, movement down (or downstream) positive. But there is another, different, set of polarities on either side of the deictic zero point. In this set, afternoon is negative and morning positive; below is negative and above positive. One can also add female and male here, for within the pair of polar opposites (and as a separate matter from the female pivotal point) female corresponds to below and is negative while male corresponds to

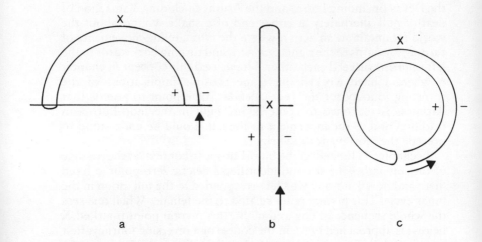

*Figure 10.1*

above and is positive.

There are two different series (figure 10.2).

1 morning : afternoon :: above : below :: male : female

2 day : night :: movement to the right : movement to the left ::
  movement down : movement up :: life : death

*Figure 10.2*

Day includes both morning and afternoon. Afternoon is negative in relation to morning but positive as forming part of the day in relation to night. The positives in one series relate only indirectly to the positives in the other and similarly with the negatives. It was realising this that led me to formulate the idea of clusters or blocks of polarities which can be distinguished from each other as lying along separate axes. My finding is that the Indo-Europeans had a system of three axes, which I have labelled A, B, C (chapter 7). The two explored here are the A and C axes and can be set out as in figure 10.3.

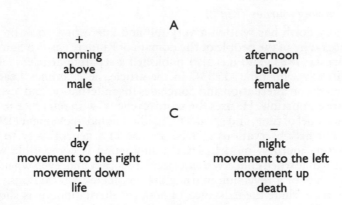

*Figure 10.3*

Louis Dumont, who is one of the scholars responsible for our growing capacity to deal with dualities within total complex systems, objected to the setting out of polarities in two columns (1979: 807-8) and R.H. Barnes has tentatively suggested that such tables should be avoided (1985: 14). I also found a single pair of columns unhelpful when working with Indo-European material, but the hypothesis of three separate axes of polarity does imply that an aspect of the totality is the occurrence of blocked sets of polarities, and any one of the sets can quite reasonably and usefully be presented in column form to show that the polarities listed belong to the same sequence. Polarities from the same sequence are correlated so that, e.g., day : night :: life : death. Polarities from different sequences, on the other hand, are not correlated but only have the weak connection that all positives on any axis are distantly related and all negatives on any axis are distantly related. In any culture where blocks of polarities lying along separate axes are involved, the exploration of the context of an individual polarity would have to take into account the fact of its membership in the block, and I suggest that the theory of axes offers one possible approach to the study of polarities in their contexts. In a first attempt at comparative study of axes, I shall now move on to a consideration of the Rindi, whose society appears both to rest on a single axis and to offer an example of the circling journey model of cyclical time.

*The circling journey (Rindi)*

Gregory Forth has written a very full and interesting book on the ethnography of the people of the domain of Rindi in eastern Sumba, Indonesia (1981) and has also published a separate article dealing specifically with time (1983). In the article, he examines Leach's comments on alternation and concludes that alternation and 'cyclicity' are compatible. He uses the word 'cyclicity' with reference to the circle model of time, and speaks of 'unidirectional' movement (1983: 75), but his observations can be applied in a general way to the 'reversing journey' model as well; alternation is compatible with cyclical time when conceived as a circular journey and, even when the two halves of a reversing journey are considered alternations, this does not exclude the existence of another alternation, as is shown above. The two major stages referred to in the following quotation are the dry and wet seasons (1983: 74):

> One indication of [the] compatibility of cyclicity and alterna-
> tion in Rindi representations is of course the fact that standard
> periods of time, such as the year, are each divided into a number
> of stages that form together a single cycle of recurrent events,
> while conterminous with this cycle is a binary partition into
> two major stages, by reference to which the passage of time
> within and between each period manifests a pattern of alterna-
> tion.

The cyclicity is 'unidirectional' and the direction is characterised as 'movement to the right'. While movement to the right in the Indo-European context is clockwise, it is anti-clockwise for the Rindi. It may be noted that the island of Sumba is south of the equator and, although Forth does not refer to the fact in connection with move-ment to the right, the sun would be seen going anti-clockwise from the east to the west through a point high in the north. The anti-clockwise 'movement to the right' is a sunwise movement, and this makes it fairly easy to relate across to the Indo-European structure although one has to remember to reverse.[2] Forth observes that, 'The Rindi also use the expression 'movement to the right' to describe the manner in which women should pass between groups in marriage, as defined by their system of asymmetric alliance' (1985: 106). In this case, too, there is alternation as well as cyclicity, with wife-givers being related to the first half of the cycle and wife-takers to the second half. Forth also refers to alternation in the case of the other cycles of time, the lunar month and day and night, commenting that 'the two major divisions of the periods in question – day and night and the waxing and waning moons – are expressly distinguished as male and female, and in various ways are associated with life and death respectively' (1981: 419).

Forth does not mention an alternation between morning and afternoon. He does, however, speak of sunrise, sunset, and midday as 'significant times of transition' (1983: 57) and so it may be that there was some contrast although it does not appear to have been stressed. This leaves us with the lunar cycle as the place where the deictic zero-point can be distinguished and we find that the period about full moon is clearly marked as a transition period, which is associated with red (1981: 208; 1983: 69). It is especially interesting to me, after casting about in the scattered Indo-European materials and finding traces of a connection between this point and the female and marriage, to find in the rich ethnographic record here that the period of full moon is connected with the transfer of the bride from wife-givers to wife-takers (1981: 205-8, 376-81). The period of transition is also clearly present in the case of the year where a special period called *wula tua* marks the interval between the dry and wet seasons. Figure 10.4 shows the year cycle as a circling journey passing through the deictic zero-point midway. Note that passage is made by way of this nodal point only once, not twice as in the case of the reversing journey. The 'circling' journey may be represented in other ways than as a circle, as, e.g. in figure 10.5 below, and it is probably the matter of whether one is approaching the nodal point repeatedly from the same direction once every cycle or approaching it twice every cycle from two different directions that may allow us to distinguish the 'circling' from the 'reversing' journey without confusion.

*Figure 10.4*

It will be recalled that in the Indo-European context the head was associated with the beginning of the light half and that the related movement began at the top of the upright body. For the Rindi, however, the upright body is not a primary image. Instead, a key image is that of the tree (1981: 32), with the base being seen as source so that movement begins from the bottom. Movement from base to top is positive and movement from top to base negative. The polarities discussed here can all be placed on a single axis, as shown below in terms of a movement from base to top followed by a movement from top to base, with the transition point, X, at the top (figure 10.5).

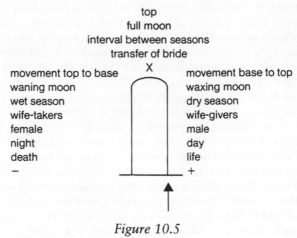

*Figure 10.5*

My impression is that the structuring can be regarded as operating mainly with a single axis of polarity, to which there may be exceptions, but exploration is at an early stage and I merely wish to open up the whole matter of axes for discussion, along with the matter of types and divisions of cyclical time, to which it appears to be closely tied.

## Hierarchical levels and encompassment

Forth (1985) discusses Dumont's idea of hierarchy as an encompassing of the opposite in connection with a particular feature of eastern Sumbanese life, the direction in which men and women wind their hair, and his discussion led me to consider his calendrical material from this point of view. It seems that the year cycle offers a hierarchy of values where movement to the right (positive) encompasses its opposite, movement to the left (negative). As shown above, the total year is seen as a movement to the right. Of the alternations it contains, the first corresponds to life and the second to death. Forth tells us that 'the Rindi state that the rule of movement to the right governs all matters connected with life' and that 'the dead are governed by the opposite rule, "movement to the left"' (1985: 105-6) and so, showing the two levels involved, we have figure 10.6.

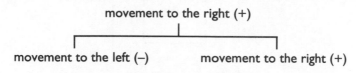

*Figure 10.6*

The function of the positive member of the polarity is shown in this instance to be more important in relation to the whole than that of the negative member, as Dumont predicts (1979: 810-2; cf. Forth 1985: 113).

This makes an interesting contrast with the Indo-European cycle with its reversing journey where the two halves are quite distinct, one having movement to the left and the other movement to the right. There is encompassing of a kind, however, and this particular type of encompassing appears to be a principle of organisation that runs right through the system (chapters 1, 2, and 5). The current interest in hierarchy gives a welcome opportunity to try to articulate its nature more fully. Taking day as an example, we find that the C axis has day as an item contrasted with night. On the A axis, morning and afternoon are contrasted and together make up day. One way of looking at the situation is that, if day is to be preserved as a separate item, there must be a means of saving it from being obliterated by morning and afternoon which cover the same period of time. And here it will be very useful to turn to N.J. Allen's discussion of possible ways of presenting encompassment in diagrams.

Dumont, as Allen says, 'pictures hierarchical opposition by drawing an outer rectangle representing simultaneously the whole and the super-ordinate element, and a concentric inner rectangle representing the subordinate element' (1985: 26; cf. Dumont 1971: 70). Allen suggests that a more satisfactory image 'starts with the square representing the totality and adds the subordinate rim *inside*'. The diagram he gives, as in figure 10.7, separates rim and central component by a dashed line, 'so as to give weight to the priority of

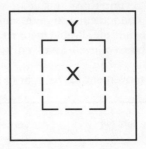

*Figure 10.7*

the whole'. 'X,' he notes, 'continues to represent both the whole universe of discourse and that which is opposed to Y', i.e. as an expression of the Rindi example above, X would correspond to movement to the right and Y to movement to the left.

Allen comments on some advantages of placing the super-ordinate element in the centre and notes (28-9) the association of the king in India with the centre. Certainly, in relation to kingship I would generally favour this central placing (cf. Rees and Rees 1961: 122-3, 133), while agreeing with Allen that 'there is room both for the encompassment and centrality representations' (27).

In some ways I think it would be more revealing to look at the space represented by Allen's diagram and verbal description in cross-section through the middle. First we have the totality (figure 10.8a) and then a subordinate rim is added inside (figure 10.8b).

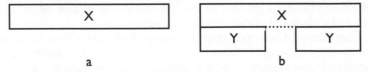

*Figure 10.8*

This is certainly a very useful way of approaching the problem of the day and the morning and afternoon (the 'rim' in this case consisting of two components) for the diagram shows an entry into the higher, encompassing level and preserves its identity by having a token representation of it visible from below. The centre need be only a single small point of entry for this to remain true. This, then, suggests that morning and afternoon do not entirely fill the day but that a small 'space' is left to allow access to the higher level. The 'space' need not be at the centre, however, although it is likely to be at a location associated with pre-eminence. It could come at the beginning of a temporal period, as in figure 10.9.

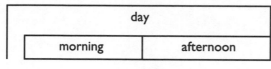

*Figure 10.9*

I have already applied a system of representation like this in my model of archaic calendar (chapter 6) although I had not seen it in quite these terms, which probably make it easier to relate this particular type of encompassment to discussion of hierarchy. In the calendar model, the points of access to the superordinate levels are the main festival days throughout the year. It should be noted that this type of encompassing is in no way confined to duality; the encompassed components in the calendar range from two to twelve in number. The polarity here is a matter of different levels, and the hierarchical duality that appears to be involved is that 'entity as a whole' is superior to and encompasses 'entity as parts'.

There are traces of the concept of the encompassing whole in both Celtic tradition and the Vedas (Rees and Rees 1961: 194-5, 200-1; Gonda 1976: 8). In the Vedas, it takes the form of what Jan Gonda calls 'the remarkable ancient practice' of adding the whole to the number of the components, and it can be suggested that figure 10.9 seen from below could give rise to a formulation like the Vedic one. One sees the parts and one glimpses, fractionally, the whole by means of the representative of it which is a whole in the guise of a part. There are three elements to be counted: the first section of the day (morning), the second section of the day (afternoon), and the day.

In the case of the reversing journey of cyclical time in the Indo-European context, movement to the left and movement to the right are on the C axis and this level encompasses the level of the A axis as shown and also the level of the B axis not discussed here. This leaves us with a duality of darkness and light; neither is the totality. In line with the general scheme of encompassment, it seems that this pair in turn is encompassed – by the female totality that is glimpsed fractionally at the zero-point.

# 11. The Place of the Hostile Twins in a Proposed Theogonic Structure

Two pairs of parallel twins have to be reckoned with in comparative mythology, as Donald Ward recognises in his study of *The Divine Twins: An Indo-European Myth in Germanic Tradition* (1968). He is concerned not with the hostile twins but with the other pair – the Aśvinic twins who occur in the Vedas as the Aśvins. They are the Greek Dioscuri, Castor and Polydeuces,[1] who appear in Roman tradition as Castor and Pollux. The Aśvinic twins are often treated simply as a pair and Dumézil has regarded them both as embodiments of his third function (fertility and prosperity). They do have distinguishing features, however, which have allowed Douglas Frame to identify them by the terms 'warrior horseman' and 'intelligent cattleman' (Frame 1978: 143). Ward has considered the material on the Aśvinic twins in relation to Dumézil's concept of the three functions of the sacred, physical force, and prosperity (Ward 1968: 20-4), and has concluded that only the 'cattleman' twin represents the third function and that the 'horseman' twin should be related to the second function (physical force). I agree that this seems probable, and shall touch on the place of each of the Aśvinic twins within the proposed pantheon later in this chapter, but I am concerned here mainly with the other pair of twins – the hostile ones. Ward is quite clear that they are a different pair from the Aśvinic twins that are the subject of his study, but he does have a brief reference to them (1968: 6-7):

> Twins often show great hostility to one another. There is even a widespread belief that twins fight within the womb. The most famous example of such hostility is the Biblical pair Esau and Jacob (*Gen.* 25: 22). A parallel example occurs in ancient Greece between Acrisios and Proitos (Apollodorus, *Bibl.* II, 2, 1). The fact that divine twins frequently exhibit contrasting natures may also contribute to the hostility between them in

mythological traditions. This theme is universal in its distribution, and is also known among Indo-European-speaking peoples (e.g., Romulus and Remus).

Rome offers familiar examples of the two pairs of twins: the Aśvinic pair, Castor and Pollux, of whom Castor is the 'horseman', and the hostile twins, Romulus and Remus. Dumézil, who does not give any special place in his system to the hostile twins and allows them to merge with the other pair, sometimes treats Romulus and Remus as Aśvinic twins (1970: 252-5; 1977: 182). To my mind, however, they provide a clear example of the hostile twins even down to the source of the hostility, which I would identify as being the question of which twin is to have the supremacy. The twins while still in the womb may 'quarrel over who shall be born first' (Leach and Fried 1950: 1135), but this is a secondary consequence of the idea that the one who is born first is the superior. The underlying motive for the quarrel is the matter of which of the twins is to be dominant, and although the quarrel does take the form of the fight within the womb it may be given other expressions. It is important to recognise that the struggle of the hostile twins is for the kingship.

This emerges quite clearly from the instances mentioned by Ward. Acrisius and Proetus were the twin sons of Lynceus, king of Argos, and his wife Aglaea, and Apollodorus (*Library* 2.2.1, trans. Frazer 1921: 1.144-5) says of them:

> These two quarrelled with each other while they were still in the womb, and when they were grown up they waged war for the kingdom.

The antagonism of the grown men is over which is to be the ruler and in parallel stories it is evident that this is also the cause of the quarrelling in the womb. In the Biblical account, when Isaac's wife, Rebekah, was pregnant with Jacob and Esau 'the children struggled together within her' and it was explained to her (*Genesis* 25: 23 AV):

> Two nations are in thy womb, and two manner of people shall be separated from thy bowels; and the one people shall be stronger than the other people; and the elder shall serve the younger.

The story is recounted more fully in Jewish legend (Ginzberg 1909-38: 1.313-7) and Louis Ginzberg notes that Esau injured Rebekah by tearing her womb and that Jacob was conceived first and 'should have been born first, but Esau threatened him that if Jacob did not grant him precedence he would kill their mother' (5.271 n.

16, 273 n. 22). There is a certain ambiguity in the situation. Jacob should have been born first, but Esau actually was born first. Implicit in the story is the concept that the 'first' son has the right to rule. Paradoxically, 'the elder shall serve the younger' because the one born second was the first to be conceived and was truly the first son. The topic of primogeniture and the right to rule is treated very explicitly in another version of the hostile twins story – the Zurvanite myth of the birth of the great gods Ohrmazd, god of light, and Ahriman, god of darkness, to the androgynous Zurvān, whose name means 'time'.[2] Zurvān promises that the first of the two sons in his/ her womb to be born will be king. One account states explicitly that Ohrmazd was next the birth passage (Zaehner 1955: 66, 433) and so would naturally be born first, and, even when this point is not made, it is clear that Zurvān anticipates that Ohrmazd will be the firstborn. However, when Ahriman hears of Zurvān's promise, he pierces the womb and Zurvān is dismayed when the black and stinking Ahriman presents himself and claims the kingship. Bound by his word, he reluctantly grants Ahriman the right to rule first for a limited period after which Ohrmazd will rule (Zaehner 1955: 68-9, 424-7).

This Iranian version is an undisguised theogony, but it is not to be expected that the myth of the twin gods would be given direct expression by the Romans who preferred to tell myth as history. Their account of the dispute over kingship is located on the site of Rome and takes place between Romulus and Remus when they propose to found the city. As told by Livy (1.6.4-7.3, trans. Foster 1919: 24-5), the story runs:

> Since the brothers were twins, and respect for their age could not determine between them, it was agreed that the gods who had those places in their protection should choose by augury who should give the new city its name, who should govern it when built. Romulus took the Palatine for his augural quarter, Remus the Aventine. Remus is said to have been the first to receive an augury, from the flight of six vultures. The omen had been already reported when twice that number appeared to Romulus. Thereupon each was saluted king by his own followers, the one party laying claim to the honour from priority, the other from the number of the birds.

There is ambiguity again, and again the king who has the lasting rule (Romulus) may have the greater claim, but the prior claim is that of the other brother (Remus). Remus is shortly afterwards killed by

Romulus or one of his supporters. It is interesting to find it suggested in studies by Jaan Puhvel (1975-6) and Bruce Lincoln (1975-6) that Remus is the equivalent of the Indian god of death, Yama, for consideration of the hostile twins narrative points to the same conclusion. The contrast between the twins, which is particularly apparent in the Iranian tradition, is that between light and life on the one hand and darkness and death on the other. Yama ('twin') was sacrificed by his brother Manu ('man'), who is a 'first king' figure, and afterwards Yama became king of the dead. Remus was killed by, or on behalf of, his twin brother Romulus, founder and first king of Rome, after contending that he had a right to be king. Both of the hostile twins have claims to kingship and it is probably an important element in the mythic structure that both actually do rule at different times or in different spheres. This element does not appear in the case of Remus in the Latin accounts of the founding of Rome but it is found in Greek in the Byzantine chronicle of John Malalas which presents Remus as ruling jointly with his brother (here given his Greek name Rōmos) both before and after his death:

> And after that Rōmos the founder of Rome was king, and his brother Rēmos. ... Which brothers conceived a hatred of one another in their kingship; and Rēmos was murdered by Rōmos, and Rōmos ruled alone.
>
> After he had killed his brother, the whole city of Rome was shaken by earthquakes, and there were civil wars during his reign. And Rōmos went to the oracle and asked, 'Why are these things happening during my kingship only?' And the Pythia said to him, 'If your brother does not sit with you on the royal throne, your city of Rome will not stop shaking nor will the people nor the war be quiet.' And having made from his brother's picture an image of his face, that is to say a golden portrait-bust of his figure, he stood this image on his throne, where he sat. And in this way did he rule thereafter, with the gold image of his brother Rēmos sitting beside him. And the shaking of the city ceased, and the strife of the people was stilled. And whatever command he might give when issuing laws, he uttered as though from himself and his brother, saying, 'We have commanded and decreed'.[3]

It is a key feature of the narrative of the hostile twins that they dispute over which of them is to be the ruler, i.e. they are royal twins, and the fact of their kingship is quite as important to their identification as the fact of their hostility. The final picture given by Malalas is of amicable joint rule by two brothers but it is noteworthy that contrast is maintained between the brothers through the opposition of their connections with the living and the dead. The twinship of the two kings is not present in every narrative tradition but the two figures can be recognised even when they are treated simply as brothers, one of whom is king of light/life and the other of darkness/ death. In the Iranian tradition, they are Ohrmazd, god of light, and Ahriman, god of darkness. In Greek tradition, they are Zeus, king of the gods, and his brother Hades, king of darkness and of the dead. In the Indian tradition of Manu and Yama, the king of the living is treated as 'first king' while his brother is god and king of the dead, and in Roman tradition also, Romulus lives among the people and rules as their king, while the rule of Remus after his death can be a spectral one only. In Egyptian tradition, the hostile twins appear as Seth, who 'was born, not in the right time or place, but bursting through with a blow, he leapt from his mother's side' (Plutarch, *Of Isis and Osiris* 12), and Horus, who is identified with the living king, the pharaoh.

It is possible now to turn to the question of the theogony and look at the place of the king and his contrasting brother. Although some of the most interesting points about the theogony are to be found in narratives that have been relegated to the by-ways of scholarship, a valuable contribution is made in one of the world's most familiar creation stories, the *Theogony* of Hesiod. We are not concerned here with the heterogeneous progeny of mother earth. Concentration on the figure of the king narrows the focus, and we are concerned only with Zeus, king of the gods, and his ancestors and siblings. The first point to note is the very obvious one that he does have ancestors. The king is not at the initiation but at the culmination of a process of creation. The primal being from whom all things proceed is Ge, the earth as goddess. She bears Uranus, the sky as god, and, after sexual intercourse with Uranus, she bears their son, Cronus, who later deposes his father. Cronus takes to wife Rhea, a daughter of Uranus and Ge, and she bears three daughters and two sons, all of whom he swallows, and finally bears Zeus, who is saved by a trick from his father and lives to overpower him and force him to disgorge his brothers and sisters (Hesiod, *Theogony* 116-87, 453-500; Apollodorus, *Library* 1.1.1-2.1). Figure 11.1 shows these relation-ships. Ge appears both as primal source and as the female in the first mating pair. The old gods who are the ancestors of Zeus are separated by a dashed line from Zeus and his siblings, who are given here in order of birth.

| Primal Source | | Ge(F) | |
| --- | --- | --- | --- |
| Old gods | Uranus(M) | | Ge(F) |
| | Cronus(M) | | Rhea(F) |
| Young gods | Hestia (F) Demeter (F) Hera (F) Hades (M) Poseidon (M) Zeus (M) | | |

*Figure 11.1*

This scheme has a marked symmetry and I would like to draw attention especially to the domestic or 'Noah's ark' model of the old gods – each male with his female. I shall suggest that this is a case where an earlier asymmetrical model has been reshaped into a symmetrical form. The symmetrical 'ark' model is also found within Egyptian tradition in the Ennead of Heliopolis of which Henri Frankfort gives the following account (1948: 182):

> Far from being an accidental combination of deities who happened to have found recognition in the city, this grouping represents a concept pregnant with deep religious significance.
>
> At its head stood the creator-sun, Atum. Then followed the divine pair whom Atum created out of himself – Shu and Tefnut, air and moisture. The children of this couple followed. They were Geb and Nut, earth and sky; and their children, Osiris and Isis, Seth and Nephthys, were the last four gods of the Ennead.
>
> There is clearly a profound difference between the last four deities and the preceding five. Atum, Shu and Tefnut, Geb and Nut represent a cosmology. Their names describe primordial elements; their interrelations imply a story of creation. The four children of Geb and Nut are not involved in this description of the universe.

Frankfort adds that 'if Horus, the living king, stood outside the Ennead, he was yet the pivot of this theological construction' and 'is sometimes called the "the tenth god"'.[4] Including the king as Horus there are ten gods in this sequence, the first part of which consists of 'the five cosmic gods', as Frankfort calls them (183), and the second part of the young gods. Figure 11.2 shows this division and also distinguishes Atum from the other cosmic gods as primal source:

| Primal source | | | Atum(M) | | |
|---|---|---|---|---|---|
| Old gods | | Shu(M) Geb(M) | | Tefnut(F) Nut(F) | |
| Young gods | Osiris (M) | Horus (M) | Seth (M) | Isis (F) | Nephthys (F) |

*Figure 11.2*

Horus is the king, but his claim was disputed by his brother Seth who claimed the kingship in a court of law and, by a first judgement, was awarded half the kingdom (Griffiths 1960: 65-74; te Velde 1967: 59-64). This judgement was later overturned, but even then, although Horus was granted the entire kingdom, the point that Seth did have some right was acknowledged by combining his symbols with those of Horus.

In the Pyramid texts of the third millennium BC, there is a reference to 'the birth of the gods, on the five epagomenal days' (Mercer 1952: 1.292, 3.883), and the calendars show that the deities tied to the five days are Osiris, Horus, Seth, Isis, and Nephthys,[5] so it appears that these five deities were from an early period regarded as a set of brothers and sisters who were born successively but so close together that they provide an illustration of the motif of multiple birth. A conception story preceding this multiple birth is told only by Plutarch in the first century AD, in *Of Isis and Osiris*, but his account gives the asymmetrical structuring of the old gods that I think likely to belong to a fundamental level of myth. The story is told in Greek, and Plutarch uses Greek names as equivalents for Egyptian gods; his Cronus seems equivalent to Geb, his Helius to Atum or Re, his Hermes to Thoth and his Rhea to Nut (Griffiths, ed., 1970: 291-4), but what matters is not the names but the structuring of the myth. Helius here plays a similar role to that of Cronus in Hesiod's *Theogony* who deliberately ate his children in order to prevent the fulfilment of a prophecy that his place would be usurped by a son greater than himself, and was only circumvented when the infant Zeus was replaced by a stone which Cronus swallowed instead of him. In this case, Helius is the 'preventer' whose action would have stopped Horus and the other young gods from being born had he not been circumvented by Hermes.

> They say that when Rhea secretly had intercourse with Cronus, Helius came to know about it and set on her a curse that she should not give birth in any month or year. Then Hermes, falling in love with the goddess, became intimate with her, and then played draughts against the Moon. He won the seventieth part of each of her illuminations, and having put together five days out of the whole of his gains, he added them to the three hundred and sixty; these five the Egyptians now call the epagomenal days and on them they celebrate the gods' birthdays. (*Of Isis and Osiris* 12, trans. Griffiths 1970: 134-5)

Plutarch adds, 'They say that Osiris and Aroueris [Horus] were the offspring of Helius, Isis of Hermes, and Typhon [Seth] and Nephthys of Cronus,' and Griffiths comments (1970: 291), 'What is problematic in Plutarch's account of Rhea is that she appears to have as many as three consorts.' It is this problematic feature which I think points to this being a fundamental form of the myth, not reduced to the symmetrical 'Noah's ark' structure found in Hesiod's *Theogony* and the Heliopolitan Ennead, but representing a single female matched with three males.

The closest parallel to the *Isis and Osiris* narrative is not a story of the gods but is told of human actors in the Irish account of the conception of Lugaid of the Red Stripes. Although it is, of course, fully recognised that mythic material may be transferred to the human plane, this particular story had not been treated as cosmogonic myth until I touched on it in a previous study (chapter 2).[6] The story of Lugaid had, however, been usefully studied by Dumézil (1973a), and so had the parallel Indian story of the conception of the four sons of Mādhavī, which I also take to be based on cosmogonic myth. These two tales suggest two different methods by which members of the final generation in the theogony are created. One is the duplication of the old gods by their children, where a young god resembles a specific parent, and the other is the combination of features derived from the old gods, where a young god has a partial resemblance to each of three fathers.

Duplication is evident in the Indian story. Mādhavī lies in turn with four men each of whom has a special virtue and gives birth in turn to four sons each of whom inherits the special virtue of his father. Dumézil and van Buitenen (see chapter 2) recognised that three of the special virtues are related to the three functions in the order: 1 ritual exactitude, 2 valour, and 3 generosity, and I have suggested that the fourth virtue, truth, although allotted here to a father and son, properly belongs to the female. A simple duplication of the four old gods, three males and one female, would give four corresponding young gods, three sons and a daughter. The whole theogony culminating in this duplication could then be seen as the binary series: 1 (goddess) > 2 (goddess plus first god) > 4 (goddess and first god plus their two sons) > 8 (goddess and three gods plus daughter and three sons).

This series, however, does not take account of the second method of creation mentioned above, that of combination, which results, as study of the Lugaid story indicates, in the birth of a king. As Dumézil has shown, Lugaid of the Red Stripes has the characteristics appropriate to each of the three functions and is a trifunctional king figure. His conception takes place when Clothru lies in the same night with her three brothers, Nar, Bres, and Lothar, and, when he is born, he has red lines round the neck and waist marking off the parts inherited from each of his three fathers, his head being like that of Nar, his body above the waist like that of Bres, and his body below the waist like that of Lothar. The strangely composed but human Lugaid has been taken to correspond to the divine Lug. The king's twin does not appear in these conception stories, but the question of the twin's parentage is of less importance in the present context than the structural point that the two young kings form a unique pair while, according to my hypothesis, each of the other young gods forms a pair with a parent of the same sex.

I understand the structure to consist of: male triad, duplicated as fathers and sons; triple female, duplicated as mother and daughter; and triple king, duplicated as king of the living and his twin brother, the king of the dead. Figure 11.3, which shows the old gods in the upper line and the young gods in the lower line, brings out these links. The placing of the female is in accord with the sequence of cosmic levels: heaven (M), atmosphere (M), earth (F), and netherworld (M), which I have discussed in chapter 2.

| Male I | Male 2 | Female | Male 3 | | |
| Male I | Male 2 | Female | Male 3 | King of Living | King of Dead |

*Figure 11.3*

It is valuable when considering theogony to keep in mind the distinction between the young gods and their predecessors, represented here by their appearance on separate lines, but there is another distinction which has to be made in relation to the hostile twins which divides the gods in a different way. The twin who is king of darkness is born first and it seems that it is only when his brother is born that light comes into being, as in the case of Ohrmazd (Zaehner 1955: 56). The other young gods, apart from the dark twin, seem all to be associated with light and, although this is a sufficiently complex matter to require separate study, it may be noted here that the Aśvinic twins are sometimes represented as stars (Ward 1968: 15-8). If we provisionally accept the young gods (with the exception of the dark twin) as gods of light, then the dark twin is connected by the absence of light to the older generations consisting of the cosmic gods. The fact that the dark twin is the first of the pair to be born probably relates his rule to the old gods, giving balanced sets of: a) the old gods under the rule of the king of darkness, and b) the young gods under the rule of the king of light. This ambiguous position of the king of the dead can be represented as in figure 11.4:

| Male I | Male 2 | Female | Male 3 | | |
|--------|--------|--------|--------|------|------|
| Male I | Male 2 | Female | Male 3 | King | King |

*Figure 11.4*

The set of five young brother gods and a young goddess is probably to be discerned expressed in human terms in the Indian epic, the *Mahābhārata*, but this again is a matter for fuller discussion than can be entered upon here. However, I shall refer to one Indian story of a mythic set of six children, taken from the *Mārkandeya Purāna* (O'Flaherty 1975: 65-70). Samjñā, the wife of Vivasvat, has five sons – Manu and Yama, the two Aśvins, and Revanta – and a daughter, Yamī, and it is clear that the Aśvinic pair is distinguished from the

Manu/Yama pair. Accordingly the Aśvinic twins should be placed in the proposed sequence of young gods not as kings but in two of the other male slots, probably those of the second and third functions as suggested above. This arrangement would give the series in figure 11.5.

| | Aśvin (horseman) | | Aśvin (cattleman) | | |
|---|---|---|---|---|---|
| Revanta | Aśvin (horseman) | Yamī | Aśvin (cattleman) | Manu | Yama |

*Figure 11.5*

Rome, in my view, provides another instance of the theogony culminating in the birth of the twins. Dumézil shows that myths were converted to pseudo-history in the Roman context (Littleton 1982: 11-2, 70-2, 90, 238-9), but he applies this insight only from the time of Romulus and Remus onwards and does not recognise the narrative before their birth as cosmogonic myth with mortals playing the roles of gods. The narrative (Livy 1.3.10-4.3; *Liber de Viris Illustribus Urbis Romae* 1.1-2 in Victor, ed. Pichlmayr 1961: 25) begins with Proca, king of Alba, who is in the position of Uranus (Male 1). He dies, leaving two sons, Amulius, who is in the role of Cronus (Male 2), and Numitor (Male 3). Amulius drives out his brother and becomes sole king. He acts as preventer when he forces Rhea Silvia, the daughter of Numitor, to become a vestal virgin so that she will not have children. However, she is impregnated by Mars and bears the twins, Romulus and Remus. In this study, the role of Mars is regarded as an interpolation, restoring the divine element that had been lost when the cosmic narrative was flattened to a human story, and Mars is accordingly omitted from figure 11.6. The twins do not have brothers and sisters, but they themselves and their human predecessors can be shown as in the following scheme.

| Proca | Amulius | Rhea Silvia | Numitor | | |
|---|---|---|---|---|---|
| | | | | Romulus | Remus |

*Figure 11.6*

Through the story of the hostile twins, it is possible to elicit a common structure with a binary base in both Indo-European and non-Indo-European material. The story is well known but, paradoxically, has not been fully studied just because it was so widespread that it did not fit in with the prevalent view that a system ought to be studied in isolation within a specific culture. While there is obviously a great deal that can only be understood in terms of a particular culture, it does not seem that the motif of the royal twins with its associated theogonic structure can very well be limited to either of the two contexts looked at here – the Indo-European or the Egyptian. It appears to have a broader base.

Indeed, the base may extend to the entire archaic old world. The question of whether the East and West are really so separate that their archaic religious systems cannot fruitfully be studied together must clearly be raised now that there are signs that the West had a complex cosmological structure which is potentially comparable with that of China.

The gods that have been studied in this chapter have been active anthropomorphic figures, or even human figures operating on the natural rather than the supernatural plane, and the problem has been to distinguish the structure that binds and defines them. In the case of China, there is no question about the existence of a powerful structuring and, if there is a basic comparability with the religious traditions already discussed, as I suggest, the problem is the reverse one of reanimating the gods that once informed the structure. Fortunately, the structural series of ten which, on the face of it, is most likely to correspond to the ten gods of the theogonies discussed above, has recently begun to show stirrings of life at the prompting of Kwang-chih Chang. This is the series of *t'ien kan* or 'heavenly stems' which are found at the period of the Shang dynasty prior to c. 1100 BC as the names of the days in a ten-day week and also as names attached to the dead.[7] Later sources indicate that half the series is yang (light) and the other half yin (dark), and, in the list of day names, yang and yin alternate so that the following set of five pairs is produced: Chia, Yi; Ping, Ting; Wu, Chi; Keng, Hsin; Jen, Kuei. Each of the elements in the Five Elements system known from the fourth or third century BC consists of one of these pairs and, in the correlation of the elements with the four quarters and the centre, the pair Wu, Chi has the central position (Needham, ed., 1954-84: 2.232, 262; de Saussure 1909: 10-1).

Given knowledge of: a) the sequence of the pairs, b) the light and dark member of each pair, and c) the pair which is distinguished from the others by its central position, one can relate the sequence to the structure shown in figure 11.4. As all these conditions are met (though b and c are on later evidence than a), the following scheme can be offered for comparison (figure 11.7).

| Yi | Ting | Hsin | Kuei | | |
|------|------|------|------|------|------|
| Chia | Ping | Keng | Jen | Wu | Chi |

*Figure 11.7*

There are some indications in the results of Chang's researches that this comparison may be on the right track. He has found from a study of the relative frequency of the *kan* names attached to rulers after their deaths that Chia and Yi (in one division) and Ting (in an alternating division) are of the greatest importance. This would tally with a model where these names are located in the slots of the hierarchically superior first and second functions. There is also the curiously anomalous position of Keng and Hsin which may ally themselves with either the Chia and Yi division or the Ting division. The equivalent slot is that of the female and we have found the female holding the balance and not being exclusively associated with any one of the functions.

The binary series found in the theogonies is, of course, thoroughly familiar in China in the progression from 1 (supreme ultimate or primal source) > 2 (heaven and earth or the two modes) > 4 (the four seasons or symbols) > 8 (the eight trigrams) (Fung 1952-3: 2.102; *I Ching*, trans. Legge, ed. Chai and Chai 1969: xliii). If it should prove possible to relate the terms of the one system to that of the other, the differences between the two treatments would no longer act as a barrier but would actually prove very valuable in providing two different perspectives on the structure.

The theogony culminating in the birth of the hostile twins could only be relevant in a society under royal rule and would serve to endorse the ideology of kingship. However, that endorsement cannot be regarded as the absolute matter it might have been if the pantheon had culminated in the birth of a single king, for the twinship points to limitation on individual royal power. In the next chapter, I explore the dual nature of the kingship in a society operating in the light of a theogonic myth in which the twin kings hold key positions. As Roy Willis has recently noted, binary conceptual systems are commonplace, the duality perhaps being grounded in the structure of the human brain (cf. Trevarthen 1984), and it is 'the decline of conceptual binarism' in Western culture that requires special explanation (Willis 1985: 209-11). It is probably this decline that accounts for the failure of mythological studies in the past either to pick up the indications of a binary structure within the Western theogonies or to give due weight to the presence of the dark brother.

# 12. Whites and Reds: the Roman Circus and Alternate Succession

In continuing this exploratory study of kingship in the context of duality, I look here at suggestions arising from comparison of features in different types of society, and indications within a particular culture. For comparison I mainly take Shang China as set against Indo-European societies, and for internal indications I examine the foundation legend and circus tradition of Rome.

The task of comparison has been simplified by the recent advance in the understanding of the relationship between certain small-scale societies and the civilisations stemming from such societies which comes from a grasp of the dualistic structure present in both. The subject is discussed fully in Maybury-Lewis and Almagor (1989) and is articulated briefly in Maybury-Lewis (1985: 19):

> Dual organization is ... a kind of world view that links the social order with the cosmic order. It is a theory of equilibrium which, if put into practice, attempts to maintain social peace by modelling it on cosmic harmony. In relatively small societies, that are not subject to the central authority of a state, the effect of dual organization is to guarantee justice, since it constrains the social system within the parameters of cosmic equilibrium. This delicate balance is threatened by state formation, unless the rulers themselves subscribe to the theory and put some form of it into practice. This, I suggest, is what happened in ancient China, ancient Egypt and the Inca empire. The absolutism of their rulers has to be seen in context. These ancient empires were organized along dualistic lines and ruled by divine kings, who linked human society with the cosmos while mediating in their persons the contending forces that could wreak havoc on earth.

Maybury-Lewis's analysis of the process of 'passage from dual organization in a tribal society to an empire ordered on dualistic principles' (1985: 19) throws a good deal of light on the question of why, when one comes at the societies through the study of dualism, each of the two different types appears to be capable of assisting in the interpretation of the other. It is only in the case of Mesopotamia, Maybury-Lewis argues, that the adaptation to a pristine civilisation was accompanied by the introduction of secular values that radically altered the picture in a way that would make comparison with a small-scale society less directly relevant. I shall be suggesting that, in the matter of alternate succession, the historical and archaeological evidence from Shang China and the indications from the myth and practice of Indo-European tribes are mutually reinforcing.

The institution of sacred kingship was common to tribes and states. Its existence did not compel a society to transform itself into a state, but it did pave the way to statehood in the case of pristine old world civilisations. Other factors are, of course, involved in the creation of a state society (Claessen and Skalnik 1978: 624-5), but sacred kingship has an important part to play, as Robert Netting points out (1972: 233):

> I would claim that on the road to statehood, society must first seek the spiritual kingdom, that essentially religious modes of focusing power are often primary in overcoming the critical structural weaknesses of stateless societies. ... The overwhelming need is not to expand existing political mechanisms (they are in certain respects radically inelastic) but literally to transcend them. The new grouping must be united, not by kinship or territory alone, but by belief, by the infinite extensibility of common symbols, shared cosmology, and the overarching unity of fears and hopes made visible in ritual. A leader who can mobilize these sentiments, who can lend concrete form to an amorphous moral community, is thereby freed from complete identification with his village or section or age group or lineage. The cultural devices for actualizing such a status are as varied as human imagination ...

I want to look at a particular cultural device for actualising the status of king in the archaic old world. It is a quite specific one with a) a cosmology resting on three axes of polarity, and b) alternate succession.

In previous chapters I have defined three axes of polarity in old world cosmology (see especially chapter 7). Now that these dualities have been set out, it is possible to move on to the study of alternation, which can usefully be thought of as duality in action. The proposed system is not a static but a mobile one, active in time, and the three axes of polarity give rise to, or at least permit, the rather intricate pattern of double or cross alternation that I shall describe.

Since I find that alternation is of central importance in Indo-European and other old world cosmologies, I am pleased to see that Rodney Needham has recently concluded that it should be given the status of a basic concept (1983). Bruce Lincoln, too, has devoted much of his study on *Myth, Cosmos, and Society: Indo-European Themes of Creation and Destruction* (1986) to exploring alternation, referring, e.g., to 'two processes alternating in a never-ending cycle' (35) and noting 'that within the common IE system of cosmological speculation there is no movement without a counter-movement' (127). His discussion may be related to earlier chapters (5, 6, and 11) where I have treated what I have defined as the C axis of polarity (light/darkness or life/death). Lincoln does not deal with alternate succession and, as will be seen, this lies along a different axis of polarity (B, wetness/dryness), although the life/death polarity is also relevant to a total view of the alternations involved.

The notion of alternate succession to the kingship is relatively unfamiliar outside anthropological circles and the procedure may therefore seem an unlikely one, but the system can be studied in the modern period working fluently in the Abron kingdom of Gyaman in West Africa. In this kingdom, there are two maximal segments, known respectively as Yakase and Zanzan. 'Every sovereign of Yakase origin must be succeeded by a sovereign of Zanzan origin [and vice versa]. ... In the sixteen royal successions since the Abron arrived in their present territory, i.e. from the end of the seventeenth century to the present time, there has only been one infraction of this rule.' (Terray 1977: 280) North-East Africa provides a number of cases of another institution relevant to this discussion, that of alternating generation sets. I will take two examples where the alternate generations are identified by colour since I will be referring to colour in the Indo-European context. Among the Turkana, there are two groupings or alternations. Every male child at birth automatically becomes a member of that of his grandfather, and is therefore in the opposite one from his father. The members of the alternations are called the Stones, who are especially associated with black ornaments, and the Leopards, who are especially associated with white ornaments (Gulliver 1958). Among the Karimojong, those belonging to one of the two alternate generation sets are referred to as yellow and wear brass ornaments, and those of the other are referred to as red and wear copper ornaments. Dyson-Hudson, in his account of the Karimojong, states that the system gives 'a sense of social continuity, of time and traditions continually recreated and relived' (Dyson-Hudson 1963: 399).

After this preamble, which indicates the basic nature of alternation and shows it in operation in society, it will not come as a surprise that a form of alternate succession can be suggested for archaic peoples in China and Europe. I am not implying that the three-axis system I am studying included alternation in quite the same form as in any of these African instances, but both alternate succession and contrasted generations seem to be essential concepts in the system, and these modern examples make it possible to flesh out the ideas by showing them in operation in the contexts of particular societies. We have to drop all the specific associations, carrying over only the concepts themselves and the conviction, derived from the particular instances, that they are effective means of social organisation.

To turn now to Shang China and the Indo-European group of peoples, the first thing to note is that there have been marked advances in understanding structure in both areas within the last fifty years or so. The big advance in China has been archaeological, the finds having been interpreted in the light of anthropological and historical examination. Study of Shang oracle bones of the second millennium BC dug up within the present century has confirmed, with only a few exceptions, the genealogy of the Shang dynasty in *Shih chi* (c. 100 BC) which had been regarded as legendary (Chang 1976; 1980: 3-7, 174-5). What appear to be the tombs of Shang kings have also been discovered in a cemetery at An-yang, with seven placed in a western sector and four in an eastern sector (Chang 1980: 111-9). Kwang-chih Chang has made the suggestion that both kinds of evidence point to a type of succession by which kings in alternate generations belonged to different moieties (Chang 1976; 1978; 1980: 165-89). The matter has been the subject of considerable debate, but I think it may be said that the main features of Chang's case are being accepted. A recent study of *The Socio-Political Systems of the Shang Dynasty* by Lin Chao is critical of Chang's work in detail, particularly as regards kinship, but accepts his main position.

The suggestion that kings in alternate generations belonged to different moieties is made possible because of having information both about succession for seventeen generations, and also about the *kan* signs which were employed as the second element of the posthumous names of kings. We need only pay particular attention at the moment to two *kan* signs of outstanding importance – Yi and Ting. The oracle bone inscriptions refer to the 'Yi door' and the 'Ting door' and do not refer to any other *kan* signs in relation to doors. This point and their greater frequency set them apart and Chang takes them as the main representatives of the two moieties. In figure 12.1 (adapted from Chao 1982: 14) I have numbered the seventeen generations of the Shang dynasty and indicated whether a generation is Yi or Ting. Chia is paired with Yi in the *kan* system and so is shown as equivalent to Yi in this context; Hsin is found in association with both Ting and Yi. It will be seen that the succession often passed to a member or members of the same generation, and that the only definite exception to the alternation of Yi and Ting in alternate generations is in generation 8 where the change occurs within the generation. This generation is also anomalous in other ways, as Chao notes (1982: 12-3), and, in indicating alternate generations by X and Y, I have treated generation 8 as if it consisted of two generations.

According to my theory of the three axes, the concept of the establishment of the cosmos in three phases or generations was common to China and the West and was early and fundamental, and, by this interpretation, Yi and Ting correspond to the Greek Uranus and Cronus and represent the first two generations. I suggest that they establish the two types of kingship and that all later kings in both traditions are either Uranus/Yi kings or Cronus/Ting kings and that the change from one to the other comes ideally with a change to a new generation. The concept of alternate succession by generation is compatible with a) a total population divided into generation sets, or b) a royal line, as among the Shang, or c) a marriage of the new king to a daughter of the former king, as in the Classical legend of Oenomaus discussed below. From the point of view of this analysis, it does not matter whether the new king comes from inside the total community, or from outside the total community as the type of stranger-king discussed by Sahlins (1981; 1985: 73-103); what is implied is that the king of a new generation is from a different category than the former king.

On the Indo-European side, the advance in understanding structure has come largely through the painstaking and imaginative researches of Georges Dumézil (see Littleton 1982). Dumézil brought

| Generations<br>(Predynastic) | | Names | Yi or Ting |
|---|---|---|---|
| | | Pao-chia | |
| | | Pao-yi | |
| | | Pao-ping | |
| | | Pao-ting | |
| | | Shih-jen | |
| | | Shih-kuei | |
| X | 1 | Ta-yi (T'ang) | Yi |
| Y | 2 | Ta-ting | Ting |
| | | P'u-ping | |
| | | Chung-jen | |
| X | 3 | Ta-chia | = Yi |
| Y | 4 | Wo-ting | Ting |
| | | Ta-keng | |
| X | 5 | Hsiao-chia | = Yi |
| | | Yüng-chi | |
| | | Ta-wu | |
| Y | 6 | Chung-ting | Ting |
| | | P'u-jen | |
| X | 7 | Ch'ien-chia | |
| | | Tsu-yi | Yi |
| Y | 8 | Tsu-hsin | = Ting (or Yi) |
| X | | Ch'iang-chia | = Yi |
| Y | 9 | Tsu-ting | Ting |
| | | Nan-keng | |
| X | 10 | Hu-chia | |
| | | P'an-keng | |
| | | Hsiao-hsin | |
| | | Hsiao-yi | Yi |
| Y | 11 | Wu-ting | Ting |
| X | 12 | Tsu-keng | |
| | | Tsu-chia | = Yi |
| Y | 13 | Lin-hsin | |
| | | K'ang-ting | Ting |
| X | 14 | Wu-yi | Yi |
| Y | 15 | Wen-wu-ting | Ting |
| X | 16 | Ti-yi | Yi |
| Y | 17 | Ti-hsin | = Ting (or Yi) |

*Figure 12.1* The Shang Dynasty

out the dual nature of Indo-European kingship and his ideas have been developed most interestingly in relation to Sparta by Bernard Sergent (1976) and have been set in a broad comparative framework by Rodney Needham (1980: 63-105; 1985). Dumézil's three functions relate to the colours white, red, and blue/black in that order, and Uranus belongs, in my interpretation, in the first slot (white) and Cronus in the second slot (red). I shall use these colours to distinguish the first-king or X generation (white) and the second-king or Y generation (red).

Dumézil's understanding of the kingship developed over time. Initially recognising two complementary aspects of sovereignty, he placed both in the first function (1948; 1977). Later he particularly developed the concept of the trifunctional king relating to all three functions (1973a). Working closely with Dumézil's ideas and material, Dubuisson has expressed the view that all three functions have complementary aspects, which he represents as A + A', B + B', and C + C'(1985: 109-10, 117). I also find that this is so, and argue that the two aspects can be referred to as light and dark (chapter 11). Dubuisson also (110-1) has explored very fully the concept of the king as the synthesis of the three functions (*la synthèse des trois fonctions*) and represents this by X(s3f). I find that the king as synthesis also has a complement in his dark twin and, in terms of Dubuisson's notation, I would accordingly say that the total scheme, excluding the female element, can be set out as: A + A', B + B', C + C', X(s3f) + X(s3f)'. Since I am using X to identify one of the alternate generations and A, B, and C to identify the three axes of polarity, I shall alter the notation and express the same scheme as: 1a + 1b, 2a + 2b, 3a + 3b, Ka + Kb, which, applied to the total scheme I have used formerly (which includes places for the goddesses not considered here), can be set out as in figure 12.2.

The complementariness in the Ka + Kb formulation is that of the king as ruling jointly with the king of the dead (see chapter 11). The

| 1b | 2b |  | 3b |  |  |
|----|----|----|----|----|----|
| 1a | 2a |  | 3a | Ka | Kb |

*Figure 12.2*

other complementariness relevant to kingship I find to be that of 1b
and 2b, the dark or 'old god' aspect of the first two functions (white
and red), the king representing alternately one or other of these two.
The Varuṇa and Mitra pair that Dumézil found expressive of the two
aspects of dual sovereignty, I would tentatively place as 1b and 2b,
one white and one red. Varuṇa is associated with ordeal by water,
whereas Mitra is associated with ordeal by fire (Boyce 1975: 34), and
this distinction may possibly accord with an alternation between
wetness and dryness in the X and Y generations. In India, the living
king (Ka) is the human counterpart of Indra (Gonda 1966: 53, 132),
and it is also possible to distinguish the king of the dead (Kb) – he is
Yama, who has been identified with Remus (Lincoln 1975-6; Puhvel
1975-6, 1987: 284-90).

In the Roman tradition, told as human history, the first king is
Proca, the second king is his son Amulius and the third king is
Romulus, with his brother Remus as king of the dead. In Greek
tradition, of course, the king of the gods is Zeus (third king, following
in succession from his grandfather, Uranus, and his father, Cronus)
and his brother Hades is king of the dead. In Scandinavian tradition,
I would place Thor as third king and, like Dumézil (1977: 196-9), I
would identify Tyr and Odin as expressing the two aspects of
sovereignty among the old gods, but would prefer to hold open the
question of which is first and which is second king for further
discussion as there is no full succession myth in this case. In the Celtic
case, on the other hand, one can see a sequence of three divine kings
quite clearly (Mac Cana 1983: 58): they are Nuada, Bres, and Lug.
I suggest, then, that the following can be placed as successions of
kings in a sequence of X, Y, X generations.

Of the triad of kings, the first two establish the alternation and
each holds total power in turn. The situation is different, though, in
the third generation, and here the story of the hostile twins has

| X | Proca | Uranus | Nuada |
| Y | Amulius | Cronus | Bres |
| X | Romulus | Zeus | Lug |

*Figure 12.3*

to be seen as the narrative equivalent of 'the differentiation of light and darkness out of the primordial twilight' as Wyatt calls it, in speaking of the activity of Indra (1986: 65). The story makes it clear that the king of the living (Ka) is the one who is expected to be king in this generation, but tells that the other twin (Kb) somehow establishes a prior claim to kingship and is given the dark part of the now divided kingdom while his brother rules the light part. It is in this third generation that the process of creating the cosmos is completed and the rule of the twin brothers forms a paradigm for human kingship, but it is only in this first generation to experience a world of light and darkness that there is a need to find a king of the dead from the same generation as the king of the living. As I understand it, from this point on each king of the living becomes the next king of the dead, so that the two kings are of alternate generations. In the case of Rome, the reign of Romulus, by one account, ends with his being torn apart and the pieces of his body being buried in the earth (see Lincoln 1986: 42-3), and this is open to interpretation as the transfer of Romulus, through sacrifice, from his royal role in life to that in death.

Rome is marked by a number of traces of a period of reversal within the year, the Saturnalia for example, and these have been set in the context of current anthropological understanding by Sahlins who notes that during periods of reversal in parts of Europe and Polynesia, as well as Africa and the Near East, the reigning monarch might be 'replaced by a mock king or superseded god of the people' (1981: 123; 1985: 92). By my reading, in the case of the archaic old world system I am studying, the anti-king does not appear out of nowhere at this time, but is one evidence of a general process of reversal of above and below. Sahlins observes in a Hawaiian context that 'men wear loin cloths on their heads' during the period of reversal (1981: 113) and similarly, I suggest, the king of the dead, who is conceived of as normally ruling below the ground, makes a visible appearance above the ground at this time. The period of reversal, thought of from the point of view of the reigning monarch, can be regarded as an interregnum, but in a total view it appears to be one of the two facets of an alternation: there are two kings throughout the year, but they take turns in ruling the two realms above or below the earth. I have found in studying the archaic year calendar that it is possible to understand its structuring as including a short period of culturally imposed 'darkness' corresponding to night (chapters 5 and 6), and I see this as the period when the king of the dead rules on earth. It should be said that I take it, as is commonly done, that the ritual pattern can be expressed at varying scales, and that the annual cycle and the life cycle are in correspondence, so that study of the year cycle can illuminate alternations whether they take place annually or by generation.

I can now present in diagram form (figure 12.4) my hypothesis about how the alternation set up in the creation myth is carried on into the historical process. Rome provides in the person of Romulus a particularly good instance of how the divine modulates into the human; a comparable figure in India is the human Manu who has Yama, god of death, as his brother. It is, I suggest, the third generation as shown in figure 12.4 that is on the borderline between gods and men so that we find here Zeus and Romulus, Indra and Manu, depending on whether the divine or the human aspect is being stressed. The first two generations of kings are unambiguously gods. Already in the first generation there is a division into above and below,[1] but the below is the place of the female. The sky god, Uranus, is first king, and his son, Cronus, also a sky god, is second king, and each rules in turn the totality of all that exists. In the third generation there is a division into darkness and light so that there are two realms ruled by two kings, who are brothers, e.g. Zeus or Romulus (Ka) and Hades or Remus (Kb), with the alternations of above and below previously noted. In generation 4, there is still a trace of the divine in that the Romulus figure (Ka) becomes the king of the dead although the king of the living can be any human. Thereafter, both kings are human, and the kings of generations 5 and 6 and all additional generations are a matter of history though continuing to enact the pattern of alternations expressed in the myth of origin.

When it comes to the question of the transfer of power from one human generation to the next, Rome remains helpful, through the tradition of chariot-racing in the circus. In this case, the Romans took their charter from an explicitly Greek legend for, although it was said that Romulus introduced chariot-racing into Rome, chariot-racing itself was held to have been instituted by Oenomaus who raced against the suitors who wished to marry his daughter and killed those whom he defeated (see Nagy 1986). He was eventually defeated and killed by Pelops who married his daughter and became king.

The charioteers in the Roman circus represented four colours – white, red, green, and blue – and both Dumézil and I have related

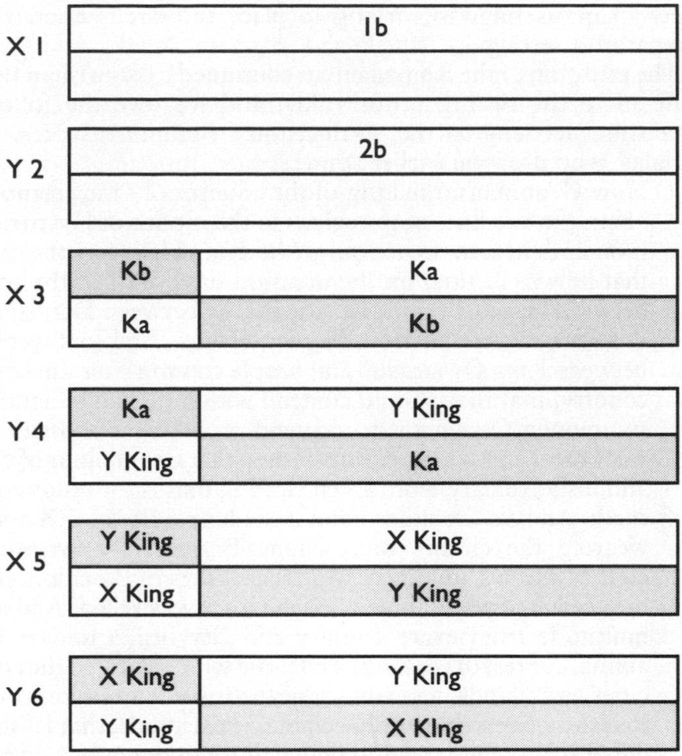

| | | |
|---|---|---|
| X 1 | 1b | |

| | | |
|---|---|---|
| Y 2 | 2b | |

| | Kb | Ka |
|---|---|---|
| X 3 | Ka | Kb |

| | Ka | Y King |
|---|---|---|
| Y 4 | Y King | Ka |

| | Y King | X King |
|---|---|---|
| X 5 | X King | Y King |

| | X King | Y King |
|---|---|---|
| Y 6 | Y King | X KIng |

*Figure 12.4*

these to Dumézil's three functions, though differing on the placing of green (see chapters 2 and 4). The four colours were paired in historical times, with dominant blue and green subsuming their partners white and red, but, since Tertullian (*De Spectaculis* 9) says that initially only white and red were worn by the charioteers, I have suggested that these may have been the dominant colours which formerly subsumed blue and green. It is for this reason that I have referred in the title of this chapter to 'whites and reds' where one would expect, in a historical context, a reference to 'blues and greens'. My view is that the contending moieties were white-and-blue and red-and-green, and that, while blue and green are more helpful in dealing directly with the circus evidence, white and red are more likely to be useful when trying to place the circus contest in a comparative setting.

The tradition of the Roman circus continued to flourish in Byzantium up to the twelfth century AD, and we owe the following interesting account to the sixth-century Byzantine writer, John Malalas, who drew on earlier sources.[2]

Now Oenomaus, the king of the country of Pisa, instituted a contest in the European regions in the month of Dystros, that is on 25th March, in honour of the Titan Helios on the ground that he was exalted on the occasion (it is said) of the contest between the earth and the sea, that is between Demeter and Poseidon, the elements subject to Helios. And lots were cast between King Oenomaus and people coming from this or that country, that they should contend with him; and when the lot summoned Oenomaus to contend on behalf of Poseidon, he wore dress consisting of blue clothes, that is the colour of water, and his adversary wore green clothes, that is the colour of the earth. And on the other hand if the lot resulted in Oenomaus wearing the clothes representing Demeter, he wore green clothes and his adversary wore the clothes of Poseidon, that is the colour of water, blue. And the loser was killed. And a vast multitude from every country and city began to watch the annual contest of the king. And those who inhabited the coastal cities and islands, and villages near the sea, and sailors, prayed that the wearers of the blue clothes (Poseidon's, that is) should win, because they augured that, if the one contending on behalf of Poseidon were defeated, there would be a dearth of all kinds of fish, and shipwrecks and violent winds. Whereas those who dwelt inland, and peasants, and all those who had to do with agriculture, prayed for the victory of the one who wore the green clothes, auguring that, if the one contending on behalf of Demeter (on behalf of the earth, that is) should lose, there would be grain-famine, and a shortage of wine and olive-oil and of other fruits. And Oenomaus conquered many adversaries for a long period of years: for he had Apsyrtus to teach him the art of chariot-driving. But Oenomaus was beaten by Pelops the Lydian and killed.

This is a very illuminating passage which aids reflection in this area of investigation. The story of Pelops and Oenomaus is elsewhere told in narrative form. Here it is given a ritual context and the chariot contest is described as a recurrent event happening every spring. The passage brings out the point that 'the loser was killed'. In the story, it could be a chance matter that Oenomaus is killed as well as being defeated in the race. Not so here; his death is obligatory. The outcome of the chariot-race determines who is to be blessed with good fortune and who is to die. The account also draws attention to the point that the king and the challenger are classified as opposites, and altogether, although it deals with specific characters, it comes remarkably close to a detached structural statement, with the following components:

> The king may be either of the land moiety, characterised by the colour green, or of the sea moiety characterised by the colour blue.

> The challenger is of the opposite moiety.

> If the king wins, the challenger is killed.

> If the challenger wins, the king is killed [and the challenger marries his daughter and becomes king].

> A win by green augurs good fortune for the land moiety and bad fortune for the sea moiety, and a win by blue augurs good fortune for the sea moiety and bad fortune for the land moiety.

> The people of both moieties are present and place their hopes on the appropriate champion.

James Vaughan, in a recent study of ritual regicide (1980), draws particular attention to the point that the killing is intimately bound up with succession, and in this passage we can see it suggested that the succession is not just personal but a matter of rotating moieties, by which the good luck goes now to one, now to the other, so that either land (the dry half) or sea (the wet half) is fortunate in turn. Similarly, in the context of Shang China, Chang notes the likelihood that the alternate generations of kings were associated with different directions, different decorative art, and different ritual practices (Chang 1976: 93-5, 103-13; 1980: 183-8). When a moiety system is involved, the entire community is liable to be directly affected by the change of kingship, one half yielding and the other half gaining the dominant position. This accession to a new role by a section of the community while another section cedes its place is familiar in societies with age class systems. Such societies today are acephalous (Bernardi 1985: xiv, 153, 157) and so cannot be fully compared with a society centred on kingship, but the friction and fighting attendant on the time of transition in these societies (Bernardi 1985: 30, 33, 60, 153) have an affinity with the ritual contest between individual opponents when these are seen as representative of halves of the community. The similarity to age class society becomes more marked when there is the difference of generation between the opponents that is indicated here by the marriage of the challenger to the daughter of the former king.

Terence Turner (1984: 360), in drawing a comparison between Dumézil's three functions and aspects of social systems in Central Brazil, has made the valuable suggestion that formal analysis 'may make possible the discovery of structural relations between functional systems of the Indo-European type, which are associated with societies in which the division of labour has developed to a point where distinct social groups can be identified on a permanent basis with particular functions (e.g. *varna*, estates, classes, etc.), and systems of a less differentiated level, in which distinct functions are associated, not with globally distinct sub-groups, but only with status-role categories of various kinds that can potentially be fulfilled by any member of society.' I doubt whether kingship in Indo-European and other archaic old world societies can be fully understood simply in terms of class, and think that looking at systems of a less differentiated level may help us to understand it, and that, in the matter of alternation, it is the structuring of societies divided into age classes that is especially helpful. As each wave of young men comes up, there is potential or actual conflict with the set of older men who hold positions of influence and who (in some cases) alone can marry (Bernardi 1985: 29, 60, 149). Eventually, the young men replace the older men, who go into retirement. The young men, now become the dominant group, are in turn under threat from the next wave of young men coming up. It seems as if this pattern is concentrated in the three-axis society in the figure of the king. The thrust behind the change of king is not then a matter of individual ambition, but the expression of the desire of a moiety for its turn. This suggestion has as corollary the idea that the first function 'priests' and second function 'warriors' are analogically related to, and may derive from, age class alternations associated respectively with white and wetness and red and dryness. The myth of origin suggests that the alternations are in some sense generation sets.

Alternate succession can occur in the form of a king of another moiety succeeding after the natural death of his predecessor, as perhaps in Shang China. However, it may well be that the other form found in the archaic old world, in which the king was subject to ritual challenge, was prior to the form depending on death in the course of nature. In this form, the power of the king is subject to severe limits, reminiscent of the limitations on individual power in the acephalous age class systems discussed by Bernardi.

The limitations imposed by the two alternations noted in this article especially in relation to Rome can be outlined as follows:

a) The king rules jointly with his dead predecessor and at ritually fixed intervals (such as every year) must give up his power to him during a period of reversal. Even when he is in power, it is likely that the moiety of the dead has a living representative that to some extent shares rule with him, although this likelihood has not been discussed here.

b) At ritually fixed intervals, the king must enter into contest with a challenger from the other moiety, and, when he loses, he is killed. This may have its positive side for the king himself, since he may accept his death and see it as having value for the community (Vaughan 1980: 123; Bloch and Parry 1982: 16) but, even if this is the case, he still does not retain power as a living ruler.

The double alternation in the hypothesised three-axis system thus emerges as a mechanism both for continually regenerating the kingship, and for ensuring that individual control of royal power remains circumscribed. It may perhaps be classed with the other means noted by Luc de Heusch (1981: 24-5; cf. 1985: 98-9) which have the effect of countering the threat of tyranny inherent in the institution of sacred kingship.

# 13. The Chinese Trigrams and Archaic Cosmology

The Chinese trigrams are well known as components of the sixty-four hexagrams of the divinatory system of the *Book of Changes* (*I Ching*) and can be studied in that context, but I am concerned here with the idea that they may have had a key place in an archaic cosmological structure that also contained the *ch'i* of heaven and the heavenly stems (*t'ien kan*). It seems as if the trigrams may actually have served as a means of notating cosmological concepts in a way that is quite unique, but is not inconceivable when one is aware of the practice of place notation which I touch on at the beginning of this chapter.

*Place notation*

In the light of the hypothesis (chapters 7, 10, and 11) that there may be a three-axis system underlying archaic old world cosmology, the threefold form of the Chinese trigram takes on particular interest as a potential carrier of three separate pieces of information. Each trigram consists of a series of three rows formed from the bottom up, or, in the case of a circular arrangement, from the inside to the outside. Each of the rows consists of either a complete line (which is yang and is associated with odd numbers) or a broken line (which is yin and is associated with even numbers), and eight trigrams can be formed from all the possible combinations of yang and yin lines. Each trigram is not simply a design as one might see it in finished form; there is the fixed order of the creation of the rows which means that each row has its place in a sequence. This has the consequence that an identical sign (say the complete line —) could potentially have a different meaning when it occurs in each of the three possible positions. In that case, the trigrams would be examples of the practice of place notation.

Place notation in the writing of numbers is very familiar in the modern world, where 272 is read as two hundreds, seven tens, and two units. In this case, it is evident that the identical figure, 2, has two different meanings depending on its position in the sequence. It is a striking fact that this familiar modern usage was also that of ancient China where decimal place-value notation was already employed, with the numbers read from left to right as above, although they were, of course, Chinese and not Arabic.[1] The design is built up cumulatively from right to left as numbers become higher, from units in the right column to tens in the second column, to hundreds in the third column, and so on to thousands, ten thousands, etc. Comparably, as already noted, the design of the trigrams is built up in a particular direction, so that it is possible to distinguish one line from another by position even when the appearance is identical. The presence of an arithmetical place-value system in the culture of ancient China makes it much more probable than it would otherwise be that the trigrams may at some stage have encapsulated meaning in a similar way.

## The ch'i

What the meaning of an individual row may consist in is strongly suggested by a recent study by A.C. Graham of the early history of Yin-Yang. Graham notes that 'the traditional cosmology as it settles into its lasting shape in the 3rd century BC is ordered by lining up all binary oppositions along a single chain, with one member Yin and the other Yang' (1986: 27), but that yin and yang, meaning shade and sunshine, earlier formed part of a set of three pairs, the other pairs being wind and rain, and dark and light, the whole group being known as the six *ch'i* (energies, influences, or forces) of heaven.[2] It is my suggestion, arising from the three-axis hypothesis, that there were formerly three chains of correspondences headed, as it were, by shade/sunshine, rain/wind, and dark/light, and that the development that took place was the falling together of the three chains into a single chain headed by shade/sunshine. The exhaustive nature of the later Yin-Yang dualism need not have been a feature of the earlier chains which could have been quite limited in scope.

Since, as Graham observes (71, 88-9), sunshine and shade were associated with heat or hot and cold, and wind and rain with dry and wet, the three pairs fit precisely into the pattern proposed for the three axes: A = hot/cold (or above/below), B = dry/wet, and C = light/ dark. In the three-axis theory, the order A, B and C is the order of creation, which may be expressed in terms of a theogony or otherwise, and it is a familiar idea in the Chinese context that the building upwards of the trigrams reflects the evolving cosmos, beginning with the separation into heaven and earth (cf. Graham 1986: 68-9; Wilhelm and Bayes 1967: 318-9), and so it can be suggested that each of the three rows of a trigram relates to the A, the B, or the C axis. It is likely that it is the initial, basic division into heaven and earth which corresponds to the dominant sunshine and shade (yang and yin) terms that were later used to refer to all dualities. This accords with the fact that hot/cold and above/below are both on the A axis in the three-axis hypothesis. There is no doubt that sunshine and shade and wind and rain are the four *ch'i* that relate to the four seasons (Graham 1986: 88). The seasons are represented by the figures formed by the first two lines of the trigrams, the images (*hsiang*), or digrams as they can be called, the bottom row of which (representing the initial division into heaven and earth) probably relates to the sunshine and shade pair, as mentioned above.

The convention of laying out the sequence of the development of yin and yang lines as a mechanical segregation table (cf. Needham, ed., 1954-84: 2 pl. XVI) with all yin lines at the left and all yang lines at the right does not (and does not claim to) reflect the actual creation process. This comes out clearly in the work of Jean Choain who gives the digrams first in the conventional segregation order and then shows them in the order of the seasons which they are said to compose (1983: 89-93), and it is of some importance, when studying cosmology, to use the latter way of representing the process of division which makes it apparent that summer and winter are the extremes of yang and yin and that the other two seasons are mixed in character. In figure 13.1, the digrams (*hsiang*) as given by Choain in relation to the seasons (which correspond to the first two lines of the trigrams in the Fu Hsi arrangement) and an equivalent table, which uses shaded boxes for yin and unshaded for yang, are shown in such a way as to reflect the cosmogonic sequence.

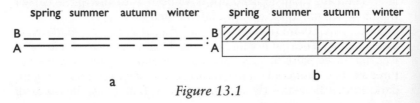

a

*Figure 13.1*

b

## The three relationships

The polarities that have been mentioned so far do not deal with society, but it is not to be expected that a symbol system would be confined solely to the level of the *ch'i* of heaven, and one social structuring that should certainly be taken into account in this connection is that of the three bonds or relationships (*san kang*),[3] which occurs in the *Ch'un-ch'iu fan-lu* (*Luxuriant Gems on the Spring and Autumn Annals*) by Tung Chung-shu (c. 179 – c. 104 BC). The three relationships are those of king and minister, father and son, and husband and wife.[4] In Tung Chung-shu's writing, the terms heaven and earth are often used together to designate the whole universe, and the heaven-earth model is reflected in the three human relationships. Authority is derived from heaven, and king, father, and husband have absolute authority over minister, son, and wife, and have also the responsibility of providing for and nourishing them. The two terms of the relationship are correlatives, and the function of either partner depends on the other. King, father, and husband, like heaven, are yang, and minister, son, and wife, like earth, are yin.

If we are using yin and yang neither in the broad 'philosophical' sense (everything is either yin or yang), nor in the limited 'etymological' sense (yang means 'sunshine' and yin 'shade') but in the proposed 'cosmological' sense which defines yin and yang as the head terms of a limited series of correlatives, then Tung Chung-shu's position should be rephrased and it should perhaps rather be said that king, father, and husband are 'plus', and minister, son, and wife 'minus',[5] and we can enquire which of the three pairs may correspond to yang and yin in the cosmological sense (lying along the A axis), and which to wind and rain (the B axis) and which to light and dark (the C axis).

In more general terms, two of the three relationships are family ones, one marital (husband/wife) and one generational (father/son), while one is political (king/minister), and we can posit the theory that there were cross-cut moieties in an earlier society along these lines,[6] but it would be premature to settle on identifications firmly before a total hypothesis concerning the archaic kinship structure associated with the three-axis system has been evolved in the light of all the available evidence. It is, however, the B axis that I have found to correlate with the proposed alternate kingship (chapter 12), and so I would suggest that the pair king/minister lies on this axis. Of the other two, I would tentatively take heaven/earth as paradigmatic for the male/female relationship and would place husband/wife on the A axis, and would therefore relate the father/son relationship to the remaining axis, C.

This would give the following correlations: A axis (or bottom row of trigram) – heaven/earth, sunshine/shade (yang/yin), hot/cold, husband/wife; B axis (or middle row of trigram) – wind/rain, dry/wet, king/minister (or alternatively minister/king, if there is alternate succession); and C axis (or top row of trigram) – light/dark, father/ son.

## *The* t'ien kan

Besides offering a method of entering into the background of Yin-Yang philosophy, Graham clears away misunderstandings about another relevant school of thought concerned with correlative thinking, that of the Five Elements or Phases, and shows how the five *hsing* as established in the third century BC developed out of previous concepts, including the *ch'i* already referred to (1986: 74-92; cf. Schwartz 1985: 356-8). As Kiyoshi Akatsuko notes (1982; 1986), it is putting the cart before the horse to interpret the ten *t'ien kan* (heavenly stems) in terms of the five *hsing*; the Five Phases are relatively late arrivals while the *kan* are fully attested in the second millennium BC (cf. Needham, ed., 1954-84: 2.264, 3.396-8). When we find the ten *kan* placed in the four directions and the centre and treated as subdivisions of the five *hsing* (Saso 1972: 54; Needham, ed., 1954-84: 2.262), it is a likely interpretation of the correspondences that the *kan* already had these places in the cosmic scheme and that the *hsing* were superimposed on them.

We know that the Shang attached great importance to the four directions, east, south, west, and north,[7] and Akatsuko (1986) suggests a connection between the *kan* and the Shang worship of the wind gods of these directions. It should be added that, if the Shang were regarding the turtle plastron they used in divination as an image of the cosmos (cf. Vandermeersch 1974: 40-1), it may be relevant that the plastron has nine plates, one placed centrally towards the front and eight round the edge (Vandermeersch 1974: 33; Keightley 1978: figs. 3, 9, 17). Since the central plate was treated as composed of right and left halves with a notional dividing line made evident through the placing of inscriptions, the plastron (potentially reflecting the cosmos) has ten components, which could correspond to the ten *kan*, and eight peripheral components, which could correspond to the eight trigrams. It has been open to question whether or not the trigrams were in existence as early as the Shang, but recent interpretations of Shang and Chou inscriptions identify the trigrams in the form of columns of odd and even numbers (Zhang Yachu and Liu Yu 1981-2). In later records, both the trigrams and the *kan* are placed in relation to the directions of horizontal space, two of the eight trigrams in each of the four directions, and two of the *kan* in each of these directions and the centre. The placing of the different trigrams in relation to the four directions varies (Legge, ed. Chai and Chai 1969: pl. III fig. 2; Saso 1972: 56, 58), while the placing of the different *kan* is fixed, and I suggest taking the fixed positions of the *kan* as a reference point.

We have evidence as early as the Shang dynasty of the *kan* as the names of a ten-day week and so in terms of time they can be seen to form an unvarying sequence, and, at a later date, there is evidence of this sequence running through the year cycle. In the Han dictionary, *Shuo-wen*, the *kan* are said to relate to the seasons and directions in the following way: Chia – east, Yi – spring, Ping – south, Ting – summer, Wu – central palace, Chi – central palace, Keng – west, Hsin – autumn, Jen – north, Kuei – winter.[8] The specific tie between the first item in each pair and a direction and the second item and a season (except in the case of the central pair) may prove helpful in further exploration of the meaning of the symbols, but, for the present purpose, the ten symbols can usefully be shown in five sections as in figure 13.2. The two bottom rows relating to the seasons have already been discussed. In giving the top row as alternately yang and yin, I am conforming the order of the trigrams to that of the *kan*, each pair of which has first a yang and then a yin partner. I would account for the absence of equivalents to the central *kan* by the theory that they are too complex to be expressed, as I suggest the peripheral ones can be (see below), in terms of the three components of the trigrams.

| spring | summer | central palace | autumn | winter |
|--------|--------|----------------|--------|--------|
| east | south | | west | north |
| Chia  Yi | Ping  Ting | Wu  Chi | Keng Hsin | Jen  Kuei |

*Figure 13.2*

## Bwe *divination*

Each of the *kan* may be seen as a god or spirit (see chapter 11), and the possible relationship between the diagrammatic form of a trigram which is produced by cumulatively marking the results of three separate divinations, on the one hand, and a spirit, on the other, may be illuminated by consideration of the *bwe* divination system of Micronesia, which has already been linked with the Chinese trigrams by William A. Lessa (1969). The basic figure here consists of two divination results that may be of any number from 1 to 4, which gives a total of sixteen basic figures ($4^2$) as against the eight of the trigrams, where there are three results which are notated by one of two signs ($2^3$). The divination is done by means of knots in a strip of palm leaf, and the whole system may be laid out by means of shells or dots in the sand in such a way that each figure takes its appropriate place in the shape of a canoe.[9] This shape represents a spirit canoe and each figure represents one of the spirits that travel in it. The spirits all have names and some have distinct roles or characteristics. For example, on Woleai Atoll, the knot combination 1/4, which can be laid out as the figure ⁞, is Bwogolimar who sits at the prow of the canoe and is the second in rank, while the combination 3/3 ( ⁞ ⁞ ) is Tagolap whose position is on the outrigger platform and who 'is the fourth most important man on the canoe and said to be the strongest physically' (Alkire 1970: 14). Each spirit has a name, a particular position in a total scheme, and a figure produced through a process of divination. If we think of the *kan* spirits as beings that are made manifest through name, location, and divination figure, we could say, for example, that one spirit has the name Keng, has the position of first of two spirits in the west, and has the divinatory combination 2/1/1,[10] which can be laid out as the figure ☰, and that another has the name Kuei, is the second of two spirits in the north, and has the combination 2/2/2, which can be laid out as the figure ☷. This would be to say that *kan* name and trigram coexist as expressions of the spirit. In a way, they would be alternatives, but it can be suggested that the trigram defines the nature of the spirit through the information given cumulatively in each of its three rows, and, if the proposed relationship is accepted, it may be convenient to identify the spirit with the *kan* name and to speak of the trigram as the definition of the *kan*.

In conclusion, then, it seems that Yin-Yang dualism and the correlative system of the Five Phases grew from the same root, and that the trigrams can help us to see the interconnections between them. We have to go back to the six *ch'i* which form three pairs, and see that each pair may correspond to a row in the trigrams. We have to be alert to the fact that there are eight possible permutations of three pairs, and that therefore the number six (seen as three pairs as in the case of the *ch'i*) and the number eight are interlocking ones and should not be forced apart as if they required separate explanation. The base, of course, is two: $2 + 2 + 2 = 6$ and $2 \times 2 \times 2 = 8$. When we look back from the Five Phases to the ten *kan*, we can again see the possible importance of two, and that eight of the *kan* may correspond to the trigrams. The nature of the remaining, central pair is a question to be answered through study of the *kan* themselves.

# 14. Transformations

An entire cosmological package has the potential to exist intact over long periods of time since each level or register is implicated in the whole and can reinforce the other registers. However, sooner or later in the course of history modifications do arise, perhaps, for example, through demographic or political change, or through an increase in 'world knowledge' (that is, information arising from observation or learned through culture contact which is not processed in terms of the system), or through some loss of understanding. It is important to realise that features of correspondence systems, like traditions in general, can be of varying historical depth. The tenacious power to retain can have the effect of allowing innovations to be readily incorporated. As W.E.A. van Beek remarks (1979: 533), force of habit 'appears to act not only as a buffer against change, but also as a mechanism for authorisation of any changes that may have happened in the past'. We have to realise when we look at the available indications of early systems that they may have features retained for millennia and may also have features that are so fresh that we will be able to observe them coming into being. There is plenty of room for development of discussion in this area, and I am putting forward for debate a view that some systems which might be thought of as totally different can be better understood when regarded as transformations of a single basic system. I shall consider Hippocratic, Platonic and Aristotelian schemes in ancient Greece and also schemes in the Jewish *Sefer Yetsira* and in Chinese tradition, all of which can be seen as arising from the same cosmological structure. In drawing connections between them, I shall pay particular attention to the important human body register in the network of analogies.

One striking thing to note about ancient Greece is the factor of individual choice in dealing with cosmological features. Different authors take different approaches, and the same author may try out different formulations in different works. Some statements are obviously incompatible with others. In treating a culture which is experimenting with combinations of the old and the new, it is, of course, out of the question to make a single general statement about its cosmological scheme relative to the human body, but the comparative approach can usefully be invoked to help us to distinguish the parts which can give information about archaic cosmology from the parts that are simply noise so far as the understanding of archaic cosmology goes. Recently, there has been a growing awareness of the importance in Indo-European cosmological thinking of the human body as a threefold structure. Bruce Lincoln, for example, provides an illustration which shows the head related to priests, thought, perception and speech, the upper body to warriors, strength, energy and courage, and the lower body to commoners, support, sexuality and appetite (1986: 143). The key structure of the threefold body is found in ancient Greece, but the threefold scheme of which it formed part gave place to a fourfold scheme (cf. Hanfmann 1951) which eventually became so dominant that the earlier importance of the threefold scheme has sometimes been lost from sight.

It is convenient to start with the clearly set out fourfold scheme in a Hippocratic treatise called *The Nature of Man* composed in the fourth century BC (Lloyd 1983: 260-71, 359), which deals with the four humours, blood, yellow bile, black bile and phlegm, and the four seasons, spring, summer, autumn and winter. The seasons of the year were considered to have different combinations of the hot and the cold, the dry and the wet, and, although actual climatic conditions could no doubt be variously interpreted, they did constrain the scheme to some extent. Lloyd notes, for example, the marked contrast 'between the heat and the drought of the Greek summer and the cold and the rain of the Greek winter' (1973: 44). The whole scheme as it relates to seasons and humours has spring and blood as hot and wet, summer and yellow bile as hot and dry, autumn and black bile as cold and dry, and winter and phlegm as cold and wet. The scheme can readily be laid out as a square diagram and Erich Schöner illustrates it in relation to the human body, as shown in figure 14.1.

There is no need to dwell on this fourfold set, so far as archaic cosmology is concerned, for it can be shown to be a transformation of a threefold set. As Walter Müri points out (1953: 28; 1971: 176;

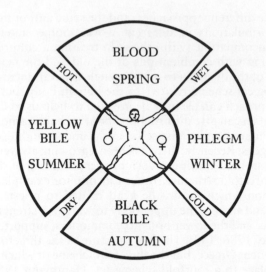

*Figure 14.1*
Fourfold scheme of the humours and seasons (after Schöner)

cf. Schöner 1964: 57-8), the four humours scheme was arrived at by subdividing bile into yellow bile and black bile, and, as he also notes, a fourth season was added to the original three of spring, summer, and winter in a similar way, with autumn emerging from the latter part of summer. Starting with spring, the new introduction in the course of the year is the third in the series; it corresponds to black bile. When autumn is not present in the series and bile is not split into two, the fourfold relationship with the body shown in figure 14.1 cannot hold; instead of top and bottom, right and left sides, we have equivalents to head and right and left sides only, as shown in figure 14.2.

Of course, the four humours have their own interest, but they do not readily put us in touch with the root concepts which relate to an entire integrated system. It is the three humours, blood, bile and phlegm – of which bile and phlegm are the contrasted pair – that can make that contact. The comparable threefold system of humours in India has been shown by Filliozat to relate to ancient cosmological concepts of wind, fire and water (1953; 1964; 1975). Speaking of the body, he says (1964: 28):

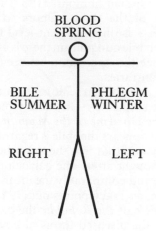

**Figure 14.2**
Threefold scheme preceding the fourfold scheme

The interplay of the three chief elements, which enter into its composition, namely, wind, fire and water, gives it life and movement. But when they are excited or when on the contrary, their action stops, disease comes in. They are, therefore, simultaneously the three elements, *tridhātus* and the three troubles, *tridoṣas* of the organism. The wind *vāta* or *vāyu*, introduces itself into the body in its own form just as in nature, that of a breath, *prāṇa*; the fire in the form of bile, *pitta*, and water in that of phlegm, *kapha* or *śleṣman*

The medical theory of the humours may well be a late derivative from earlier cosmology but one can see in the Indian case how a humoral scheme can be integrated into a whole way of organising concepts. In a recent subtle study, Zimmerman shows how it operates in relation to cultural constructs which embrace such areas as geographical divisions and the classification of animals. Important in this ordering is the idea of extremes and a mean. It is this that can be considered to have been present in the early Greek (and Indo-European) three-season pattern, with summer and winter as extremes and spring as intermediate between them. Similarly, Zimmerman comments on the Indian humours (1987: 146):

> Wind, the first of the three humors to be cited – in the traditional order – is on a different level from the other two humors. Between bile and phlegm there is a symmetry. Bile and phlegm symbolize the opposition between fire and water, an opposition of contraries.

In Plato, bile (hot) and phlegm (cold) are contrasted in a number of contexts which are studied by Theodore James Tracy in *Physiological Theory and the Doctrine of the Mean in Plato and Aristotle* (1969: 123-36), and it appears that bile is regarded as the excess and phlegm the deficiency and that it is the mean between them that is highly regarded and sought after. The balancing of the humours in the body is desirable, and comparably the mean is also sought after in the political sphere, as Tracy points out (131):

> In the eighth book of the *Republic* the dangerous proletariat that appears in the diseased forms of government, oligarchy and democracy, and is the cause of their change to a worse, is compared to bile and phlegm in the diseased body (564B), being divided into two groups, the *active* trouble-makers, the criminals or 'drones with stings', and the great, *passive* mob, the beggars or 'stingless drones'. This confirms the notion that Plato regarded phlegm and bile as morbid humors with opposite properties, bile being compared to the active, exciting, 'stinging' element, while phlegm is like the great, inert mass that swells the body politic, sapping its resources and serving no useful function.

In the *Phaedrus* (246, 253), Plato shows the ruling part of the tripartite soul, compared to the charioteer, controlling the fiery right-hand horse and the lethargic left-hand horse. The horses are the spirited and appetitive parts and these are evidently in correspondence with bile and phlegm. The controlling head is the mean, while the right side is the spirited and the left side the appetitive. The three parts of the soul are also related to the three parts of the vertical body quite explicitly in Plato (*Republic* 588-9, *Timaeus* 69A-71A), so that the head is seen to be the rational part, the upper body the spirited part, and the lower body the appetitive part. Figure 14.3 illustrates the alternative ways of expressing the threefold division in terms of the body.

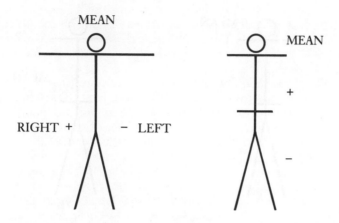

*Figure 14.3*
Two expressions of the extremes and the mean in Plato

As I have discussed in chapter 3, the three parts of the soul are related to the three notes of music, the high (*nete*), the low (*hypate*) and the controlling middle note (*mese*), so it appears that one can say that the right side corresponds to *nete*, the left to *hypate* and the head to *mese* in this schema. As regards the division of the vertical, Plato does not say which of the three harmonious notes corresponds to which part of the body, and, although Plutarch did interpret the correspondences as they are shown here, and this seems to be Plato's meaning, there was an alternative school of thought which held that the high note corresponded to the head, the low note to the lower body, and the middle note to the upper body. There is an evident reason for this alignment in the bodily position of the three components, but it marks a fresh departure – a transformation which we can explore.

In the Hebrew *Sefer Yetsira* of the second or third century AD, which was subject to influence from non-Jewish streams of thought, the letters of the Hebrew alphabet structured a correspondence system.[1] The three double letters, Shin, Alef and Mem, correspond to the head, upper body (chest) and lower body (abdomen), and head and lower body are given as the extremes of the positive and the negative, with Alef (upper body) being the balancing component. If

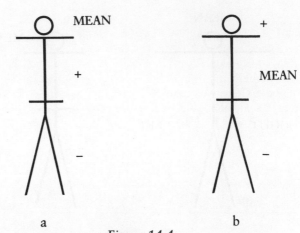

*Figure 14.4*
(a) The mean in Plato and (b) a transformation

there is influence from Greek thought here, there has been a transformation in either a Greek or a Jewish context from the Platonic triadic structure to a new threefold schema, which switches round the sets of correspondences that relate respectively to the head and heaven and to the upper body and air (between earth and heaven).

The following passages in the *Sefer Yetsira* give the relevant correspondences (Hayman 1986: 30-1, cf. 37; paras. 23, 25, 28-30, 36):

> Three matrices: Alef, Mem, Shin. Their basis is the scale of acquittal and the scale of guilt, and the language of law holds the balance between them.
>
> Three matrices: Alef, Mem, Shin. The offspring of the heavens – fire; the offspring of air – aether; the offspring of earth – water; fire above, water below, and aether is the balancing item.
>
> Three matrices – Alef, Mem, Shin – in the universe: aether, water and fire. Heaven was created first from fire, and earth was created from water, and air was created from aether, holding the balance between them.
>
> Three matrices – Alef, Mem, Shin – in the year: fire, water and aether. Heat was created from fire, cold was created from

water, and humidity from aether holding the balance between them.

Three matrices – Alef, Mem, Shin – in mankind. The head was created from fire, the belly from water, and the chest from aether holding the balance between them.

Three matrices: Alef, Mem, Shin. There was formed with Alef: aether, air, humidity, the chest, law, and language (the tongue). There was formed with Mem: earth, cold, the belly, and the scale of acquittal. There was formed with Shin: heaven, heat, the head, and the scale of guilt. This is Alef, Mem, Shin.

If this structure in the *Sefer Yetsira* is to be related to the scheme in Plato and Indo-European cosmology, it is necessary to take account in any comparison of the different use of the two upper sections of the body, the mean or balancing component being the upper body in one case (cf. figure 14.4b) and the head in the other (cf. figure 14.4a).

It is not possible to show here that the *Sefer Yetsira* scheme was historically the result of an adaptation of the Platonic type of scheme, but it is possible to say that it can be expressed as a transformation of that scheme and could conceivably have been derived from it. If so, an influence that could have aided the transition may have come from the Aristotelian sequence of the elements, for the *Sefer Yetsira* has the same sequence except that its scheme has only three slots to contain the four components of fire, air, water and earth: 1 fire, 2 air, and 3 water and earth. It will be of some interest to turn now to the Aristotelian sequence of the four elements and observe how it is tied to the concept of a spherical earth surrounded by concentric spheres which was developed by Eudoxus in the early fourth century BC.

It is remarkably easy to see Aristotle, a great figure in the history of science, as the heir of a pre-scientific cosmological tradition. He was intent on fitting everything into a single scheme of knowledge, and, although his main strength was in the biological sciences, he also dealt with questions in the fields of physics and astronomy. His view of the cosmos is succinctly stated by Margaret J. Osler and J. Brookes Spencer (1985: 835):

> [Aristotle] considered the cosmos to be divided into two qualitatively different realms, governed by two different kinds of laws. In the terrestrial realm, within the sphere of the Moon, rectilinear up and down motion is characteristic. Heavy bodies, by their nature, seek the centre and tend to move downward in

a natural motion. It is unnatural for a heavy body to move up, and such unnatural or violent motion requires an external cause. Light bodies, in direct contrast, move naturally upward. In the celestial realm, uniform circular motion is natural, thus producing the motions of the heavenly bodies.

The four elements of fire, air, water and earth belong to the sublunary world. Two, fire and air, of which fire is the extreme, are light and incline upwards, and two, earth and water, of which earth is the extreme, are heavy and incline downwards. It is because of its weight that earth is central in the cosmos (Dijksterhuis 1961: 32-4). In this geocentric world view the central spherical earth is surrounded by a large number of other spheres, of which the ones nearest earth are those of water, air, and fire in that order (Dijksterhuis 1961: 22-5, 33-4; Dicks 1970: 203). Because of the correlation in Aristotle's sublunary world between moving towards the centre and moving down it is possible to diagram his scheme of the elements either concentrically or as a series of layers, as in figure 14.5.

We have already looked at a fourfold scheme of the seasons and humours in *The Nature of Man* and can now compare that with the Aristotelian scheme of the elements which are made up of the same

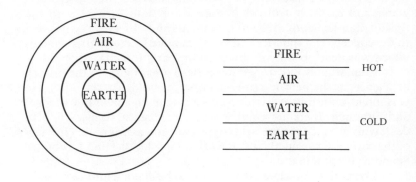

*Figure 14.5*
The sequence of the elements in Aristotle

qualities. Fire is hot and dry, air hot and wet, water cold and wet, and earth cold and dry, and the two hot and light elements belong to a higher level than the two cold and heavy elements. Spring and summer are the hot pair and can be correlated with the high level, while autumn and winter are the cold pair and can be correlated with the low level. However, Aristotle does not equate the elements with the seasons, and it would be hard to do so since hot and dry, which relate to his first (uppermost) element, fire, are clearly applicable to summer, which is the second item in the 'hot' pair of the seasons.[2] The sequences of the elements and the seasons, though spoken of in terms of the same qualities, are incompatible. If the qualities are lettered in the order in which they occur in the seasons sequence – hot and wet (A), hot and dry (B), cold and dry (C) and cold and wet (D) – in the elements sequence the letters occur in the order B, A, D, C, with a reversal within both the upper and the lower pair.

Since the sequence of the elements accords with the new theory of a central earth in a spherical universe, it seems that, if we want to understand the earlier cosmology, we are better to take the sequence of seasons as a guide. As already indicated, there were formerly three seasons, spring, summer and winter corresponding to A, B, and D in this scheme. By my theory (see chapters 2 and 9) the whole is represented at point C, which is waist level in the upright body and the level of the earth in a universe correlated with the upright body. It is interesting to see how the concept of a central earth could have been carried over from one cosmology to the other,[3] a major reconstruction being entailed by the notion of an earth in the centre of global space as against earth as the centre in a layered universe (see figure 14.6).

Although outmoded concepts concerning the vertical body and the layered universe had apparently to be revised, Aristotle could and did retain a traditional world view in the matter of relating the body to horizontal space, as we shall see below.[4]

When the vertical human body is concerned, there is no question about its placing; unless there is deliberate reference to reversal the head is at the top and the feet at the bottom. In the next suggested transformation, however, I shall be dealing with the horizontal and here it is of crucial importance to know in which direction the person is facing. It will be worth while to pause and look at the general question of body orientation in some detail.

Classification in connection with space has been treated in a recent

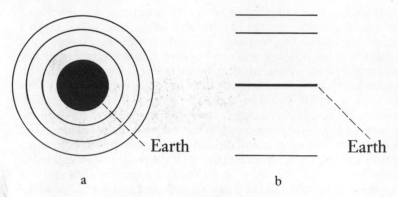

*Figure 14.6*
Central earth (a) in Aristotle, and (b) in a layered cosmos based
on the analogy of the human body

article by Roy Ellen who comments (1986: 27) that there are two
basic classificatory logics: 'One is an order relative, dependent and
responsive to humans; the other independent of the location and
orientation of people.' The first order is deictic (dependent on ego)
and, of course, human beings move around and can face in any
direction. In the study of cosmology we need to explore a middle
ground to show how space may be ordered deictically and yet have
fixed locations, and I believe it will be useful to introduce the terms
'canonical direction of facing' and 'canonical orientation' to express
the direction of facing which accords with cosmological structure.
This direction may be a world direction, such as a particular cardinal
point, or may relate to a major geographical feature, such as a river.
A specific culture may not have a canonical orientation, and a culture
that does have one may or may not encode this in its language. To
move right away from the contexts of the comparison I am about to
make, an instance may be taken from the Foi people of Papua New
Guinea. Their canonical orientation is encoded in the language and
is taken from a river that runs through their territory. The river runs
from west to east and the canonical direction of facing is upstream,
which, linguistically, gives rise to the expression *ki'u ta'i* ('backside
downstream'). The human being, of course, can be facing any way,
but canonically faces upstream and west, a direction that is associated with the male. A bush house, which may in actuality face in any

direction, is also canonically oriented upstream so that 'a man always speaks of the woman's half of the house as the east or downstream side' (Weiner 1988: 46-8, 23-4, 298-9 n. 7).

An orientation towards the east is traceable in Hebrew and Arabic (Chelhod 1973: 246-7), and is also embedded in a number of Indo-European languages, for example, Gaelic, where south and right hand are both expressed by the word *deas* (cf. Pokorny 1959-69: 1.190). When cases occur of a different ritual direction of facing within the culture areas where these languages are current, we have to look for special explanations, such as deliberate ritual reversal; the basic form is most likely to be the one that is expressed or latent in the language. There is one matter which can give rise to confusion which it may be as well to discuss here. The canonical direction of facing, like 'in front' in general, is positive. If the canonical direction of facing is east, this can clearly lead to a valuing of the east, but it may also lead to a valuing of the west, as being the direction in which the person facing the east is placed relative to the easterly point that is looked at. Once one knows the canonical orientation of a culture, one can begin to understand variations of this nature. The norm for the Hebrew, Arabic and Indo-European language areas is east-facing.

Ellen speaks of the body as 'a template providing a series of contrasts' and notes that 'the binary oppositions which we link to fixed points in the universe are themselves ultimately modelled on the human body, in the anatomical symmetries which allow us to divide the body into halves' (1986: 27). For Aristotle, 'above is more honourable than below, and front than back, and right than left' (*Progression of Animals* 706b 12f.; Lloyd 1973: 174). As we have already seen, in the vertical division of space, the two hot elements are above and the two cold elements below; the positives are 'hot' and 'above'. On the horizontal, Aristotle employs a division into halves when he links east with south and west with north, noting that an east wind is regarded as a kind of south wind and a west wind as a kind of north wind (*Politics* 1290a; *Meteorology* 364a). He associates east and south with the hot, so that the positive half is 'east-and-south'. In the human body, on the horizontal plane, the honourable half is 'in front-and-right'. Correlating the body with the universe shows Aristotle following the Indo-European norm of an orientation to the east.

By contrast, Chinese culture has been noted for its orientation to the south. The state was embodied in the ruler who, as John Major

154 Emily Lyle

puts it (1984: 153), 'ritually "faced south"; that is, he symbolically occupied the position of the Pole Star at the unrotating pivot of the universe'. It is interesting to observe, though, how far the Indo-European and Chinese cultures had comparable cosmologies for they are not totally distinct.

The two cultures are alike in possessing a strongly marked canonical orientation of one kind or another integrated into a complex cosmological structure in a society under royal rule. They are also alike in relating spring and summer to east and south, and correlating them with the upper, and hot, half.

They are unlike in two major respects relating to orientation. One is that the Chinese give pre-eminence to the left hand, left being yang and right yin (cf. Granet 1973: 45, 50–1; Graham 1986: 27–8), while the Indo-Europeans, like most other peoples, give pre-eminence to the right hand. The other is that, as noted above, the canonical direction of facing for the Chinese is south while for the Indo-Europeans it is east (see figure 14.7).

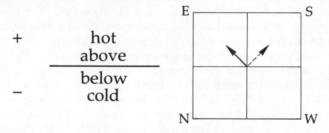

*Figure 14.7*
Canonical orientation: Indo-European (undashed line) and Chinese (dashed line)

It is in connection with China that I wish to introduce a final example of a possible transformation, one that would have had very far-reaching effects, but that is actually very simple conceptually. The anomalous privileging of the left hand can be accounted for if, in a cosmology where east and south were positively valued, there was a shift from orientation to the east to orientation to the south. After such a shift, positive value could well have been transferred from the direction, east, to the hand positioned in that direction, the left. Conversely, of course, the relatively negative value attributed to the west would have been transferred to the right hand. Where one part of a correspondence system is altered, adjustments are required elsewhere, but the form the adjustments may take is not predictable. For example, a shift from orientation to the east to orientation to the south might have led to the west acquiring superior value because it had become the right-hand side, but this possibility is clearly irrelevant in the Chinese context.

As, in general, I find that old world cosmologies form a continuum, I am ready to entertain the possibility that there may once have been a common orientation, and that the orientation of one or other of the two cultures under discussion has undergone transformation. If so, it seems probable that the shift was from east to south since this shift would supply a cosmological explanation for the unusual feature of the privileging of the left hand in China.

A range of new opportunities for the study of the old world's remote past is opened up through the cosmological approach taken in this book which suggests that we can correlate the systems of East and West once we allow both for the processes of disintegration that were typical of the West, and also for the possibility of sweeping reformulations where cosmological thinking remained strong, as in the East. When a cosmology has been fragmented, we can examine the pieces and try to see the shape of the whole, and, when one has remained basically intact, we can try to assess the effects of change over the course of time. Since archaic cosmology, though now far in the past, was the matrix in which our current world views were formed, we may well wish and need to take every possible opportunity to gain an understanding of it and to learn what we can of ourselves and others from the perspective it offers.

# Notes

## 1 Cosmos and Indo-European Folktales

1 This chapter was given as a paper in August 1979 at the Seventh Congress of the International Society for Folk-Narrative Research, held in Edinburgh.

2 For a recent development of this position, see Allen 1987.

3 The triad with three elements corresponds to the 'closed triad' and the triad with four elements to the 'open triad' as defined in Syrkine and Toporov 1968. I am grateful to Dr Mihály Hoppál for mentioning to me after I gave my paper that Russian writers on mythological semiotics, including the authors of this article, had been taking an interest in ideas like those I was discussing. See also Malamoud 1981 on a three plus one framework.

4 Dumézil 1954: 45-61. See also Turner 1967, and Berlin and Kay 1969.

5 This was written before the distinction emerged between the female whole at the pivotal, central point of the year (see chapters 2 and 9) and the period of darkness at the opening of the year (see chapters 5 and 6).

6 Ward 1970, especially 137. Cf. Mac Cana 1983: 25-8.

7 Meid, ed., 1970: 35, and English translation in Byrne and Dillon 1937: 4.

## 2 Dumézil's Three Functions and Indo-European Cosmic Structure

1 Littleton 1982: 217, cf. 101. See also Dumézil 1930: 109-30; Benveniste 1932: 117-34, and 1938: 529-49.

2 Dumézil 1977: 39 and references in n. 2; see also 252-4, 261-5.

3 Rees and Rees 1961: 81-204, part 2, 'The World of Meaning', especially 111-3, 122-33.

4 Muni, trans. Ghosh: 1950-61: 1, part 1: 25. The passage is

referred to in Keith 1924: 359, and Beck 1969: 559. For another instance of this ordering, see Rees and Rees 1961: 131-2, 376.

5 Dumézil 1958: 26, and 98, n. to paragraph 20.

6 Cedrenus, ed. Bekker 1838: 257-9. When I wrote this article I was much less familiar with circus material than I later became. I referred to Cedrenus, a writer of the twelfth century AD, since he had been mentioned by Dumézil, but later found that Cedrenus's ultimate source was the sixth-century chronicle of John Malalas (see Malalas, trans. Jeffreys, Jeffreys and Scott 1986: xxxvi). Both writers refer to Romulus dividing the city of Rome into four 'factions' (Greek μέρη). The word μέρη has the general meaning of 'sections' and appears to apply to 'parts' of the universe in Malalas, ed. Dindorf 1831: 175-6, but it would have been wiser not to give it simply the sense of 'parts' here, for, although a conceptual connection between faction colours and city quarters seems not improbable (cf. the material on fourfold division in Rome in Müller 1961), it cannot be directly proved from this passage.

7 Zaehner 1955: 31, 449-50; Ghirshman 1962: 57-69; and Dumézil 1941: 62-3.

8 This may be a matter of bias in looking at the material. The opposition of the Chinese yin (cold, dark, femaleness) and yang (heat, light, maleness) is familiar as an instance of division in halves (see Needham, ed., 1954-84: 2.273-8), but on closer inspection, it becomes evident that one particular quarter is especially identified with the female, the oppositions of the quarters being: chief (south) and vassal (north), male (east) and female (west); see Granet 1973.

9 Nicolas 1966: 83, 88, 98-101, and 1968: 570-2, 603-4.

10 Benveniste 1973: 427. See also Dumézil 1947: 115-58, and 1970: 237-40. Dumézil relates the ram to the first function and the bull to the second; in giving the reverse relationships, I follow Benveniste.

11 Caland 1969: 53-4; 16.27.7-12. The paired opposites among the animals are clear, but the directions can be interpreted differently. The most straightforward equation with the position of the functions already employed is given by showing the animals, as is done here, according to the directions they are facing (bull, east; ram, south; horse, west; he-goat, north).

12 Besides its special virtue, each class also has the virtue or virtues

of the class or classes below it in the hierarchy; see Murphy 1951: 18-20.

13 *Mahābhārata*, trans. van Buitenen 1973-: 3.410; the story of Gālava is at 3.395-415.

14 Physical division into four, but into the four castes without any reference to a female element, is found in *Rigveda* 10.90.11-2; see Lincoln 1975-6: 127.

15 An account of the division of an androgyne in the *Bṛhadaraṇyaka Upaniṣad* is set in the context of other creation myths in O'Flaherty 1975: 34-5.

16 Cf. the link between human position and direction in space for which there is linguistic evidence; see, e.g., Pokorny 1959-69: 1.190, '*deks-*' with the senses 'right' and 'south'. For the connection of what is in front and to the right with the male, see Lloyd 1966: 49-53.

17 See, e.g., the offerings made by the Greeks to the chthonians discussed in Guthrie 1950: 220-2.

18 Littleton 1982: 12, 71; Dumézil 1973b: 3-25, and Dumézil 1970: 60-78, 164.

19 Forsberg 1943; Bosch 1960: 57-64, 87-8; and Rees and Rees 1961: 83-94, 100-3.

20 It appears that division among the three groups in society is made according to these proportions in the Iranian account of the preparation by Yima of an underground enclosure; see Benveniste 1932: 119-21. Cf. Dumézil 1973a: 4-7.

### 3 'King Orpheus' and the Harmony of the Seasons

1 The ballad has no known traditional title. Child named it 'King Orfeo' after the Middle English romance which has been called both *King Orfeo* and *Sir Orfeo*, but it seems more appropriate now to name it after the recently-discovered Scottish romance, *King Orphius*, to which it is more closely related. Since there is no reason to retain the romance's sixteenth-century spelling for the ballad, I have named it 'King Orpheus'.

2 Fox, ed., 1981: 139. For background, see MacQueen 1976, and Meyer-Baer 1970.

3 West 1971: 217. West refers to Burkert 1962: 333-5. In the English translation now available (Burkert 1972) the equivalent pages are 355-7.

4 These lines are from the earliest known version of the ballad,

which was recorded by Bruce Sutherland at Gloup, Orkney, in 1865, published in *The Shetland News* in 1894, and reprinted in Shuldham-Shaw 1976. Other traditional versions may be found in Child 1882-98: 1.217, and Bronson 1959-72: 1.275. The precise meaning of 'gaber' is uncertain since the word does not occur elsewhere.

5 Stewart 1973: 6, lines 81-6. In this quotation, abbreviations have been expanded, punctuation and capitalisation added, and the letters 'y' and 'v' replaced by 'th' and 'w'.

6 The quotations are of paragraphs 73 and 164 as translated by Whitley Stokes (1891: 81, 109) except that I have substituted 'joy' for 'joyance' and 'Dagda' for 'Dagdae'. The word *cruit*, translated by Stokes as 'harp', could apply to either the harp or the lyre. The earliest iconographical evidence for the fully-framed triangular harp is of the eighth or ninth century AD, and the stringed instrument in use among the Celts at an earlier period was probably a U-shaped lyre (see Rimmer 1969: 9-13, 22; and Bannerman, forthcoming).

7 Farmer 1925-6: 99-100, quoting Al-Kindi. Farmer shows that melodic modes were also associated with the different strings in Farmer 1933: 8-15, 38-9.

8 The quotation is from O'Curry 1873: 3.223-4. For a complete version of *The Lay of Caoilte's Urn* with translation and notes, see MacNeill 1908: 38-45, 140-9, and Murphy 1953: 36-40, 274. For the playing of the three types of music on a *timpan* with three strings, see O Daly, ed., 1975: 40-1.

9 Diodorus of Sicily, *The Library of History* 1.16.1-2, trans. Oldfather 1933: 53. Anne Burton notes (1972: 17, 78-9) that chapter 16, 'which discusses Hermes or the Egyptian Thoth, suggests an equal mixture of ideas' from Greece and Egypt and that the god's association with music 'must come from the Greek tradition'. Cf. the harmony of the three seasons played by Apollo in Orphic Hymn 34, where winter is related to the low note (*hypate*), summer to the high note (*nete*), and spring to 'the Dorian'; see Quandt, ed. 1962: 27, and Athanassakis, trans. 1977: 48-9, 124.

10 'De Musica', 1.2 and 1.20; Boethius, ed. Friedlein 1966 [1867]: 187-9, 205-7. On Boethius's writings on music, see Caldwell 1981 and Chadwick 1981.

11 I.e. C, F, C' in the key of C.

12 *Moralia*, ed. Babbitt and others 1927-69: vol. 13 pt. 1, p. 331, 'On the Generation of the Soul in the Timaeus' 1029A and note *e*; cf. *Moralia*, vol. 9, p. 277, 'Table-Talk' IX.14, 745.

13 *Republic* 414-5; trans. Lee 1955: 159-61. For the divisions of the body, see 588-9 (Lee 1955: 365-7), and Cornford 1937: 279-86 (on *Timaeus* 69A-71A).

14 See Plutarch, 'Platonic Questions' IX and notes (*Moralia* vol. 13 pt. 1, pp. 91-103).

15 'Platonic Questions' IX, 1009A (*Moralia* vol. 13 pt. 1, pp. 100-1). I have adapted the translation to give the Greek names of the notes and the term 'the spirited'.

16 *Republic* 398-9. On the influence of the Phrygian and Dorian modes cf. Aristotle, *Politics*, 1340b, trans. Barker 1946: 344: 'Another mode is specially calculated to produce a moderate and collected temper; this is held to be the peculiar power of the Dorian mode, while the Phrygian mode is held to give inspiration and fire.' These modes are not, of course, the same as the later church modes known by these names; see Kolinski 1974.

17 Plato, *Republic Book X*, ed. Ferguson 1957: 69-70, 93; Radhakrishnan 1940: 147-9; Urwick 1920: 15-41.

18 Gonda 1976: 96; *Śatapatha Brāhmaṇa* 12.9.1.7.

19 Blom, ed., 1954: 4.456-60 'Indian Music'; cf. Wellesz, ed., 1957: 199-201, and Gosvami 1957: 4-7.

### 4 The Circus as Cosmos

1 Malalas, ed. Dindorf 1831: 175. Charax of Pergamum has been dated to the second century AD; see Cameron 1976: 64-5. Most of the quotations in this chapter appear in the original Greek and Latin in this study as published in *Latomus* in 1984. I am most grateful to Dr R.C. McCail of the Department of Classics at the University of Edinburgh for his translation of the passages from Malalas and Lydus in this chapter and in chapters 11 and 12 and for his ready help with the Byzantine sources.

2 See now the comprehensive study by John H. Humphrey (1986).

3 Lydus, *De Mensibus*, ed. Wuensch 1898: 1.12; cf. Isidore of Seville, *Etymologia* 18.31, and line 14 in the anonymous *De Circensibus*, ed. Riese 1869: 1.197.

4 *In laudem Iustini Augusti minoris* (*In Praise of Justin II*), 1.314-33, trans. Averil Cameron 1976: 93.

5 In my study as published in *Latomus*, I used the word *spina* but

have changed this to 'barrier' in the light of Humphrey's usage; see 1986: 175-6, where he comments: 'The term *spina*, I believe, referred particularly to the narrow side walls on which statues were placed.'

6 Dagron 1974: 331, n. 6; *De Circensibus* line 12, ed. Riese 1869: 1.197; cf. Cassiodorus, *Variae Epistolae* 3.51; Isidore of Seville, *Etymologia* 18.30; Malalas, ed. Dindorf 1831: 175.

7 Corippus, *In laudem Iustini Augusti minoris* 1.322-3; Lydus, *De Mensibus*, ed. Wuensch 1898: 4.30.

8 Philo of Alexandria, *De Aeternitate Mundi* 58, trans. Harris 1976: 61. The comparison with the seasons is in *De Specialibus Legibus* 4.233-5, trans. Colson and Whitaker 1929-62: 8.150-3.

9 Harris 1972: 154; cf. *Iliad* 23.262-533. In this study as published in *Latomus* in 1984, I suggested that the triple form found in the Roman turning-post might already be identified in the *Iliad* in the ensemble of tree-trunk and two white stones. I now see from Humphrey's recent book (1986: 255-6, 659 nn. 155-6) and an earlier article he cites (Quinn-Schofield 1968) that this suggestion had already been made in the sixteenth century by Giovanni Argoli, and that it is based on a misunderstanding of Homer that has found its way into many translations of the *Iliad*. Instead of the picture conjured up by Harris's description in the passage quoted, we should envisage a length of timber about a fathom above the ground supported by a white stone at either end to form a gateway or *dodoka*. This bears no resemblance to the design of the Roman turning-post.

10 Michels 1967: 19-21, 130-2. For the division of the waxing half employed in figures 4.4b and 4.5, see chapter 6.

11 Michels 1967: 89, 167, 192; see also Colson 1926.

12 Nilsson 1920, and Richardson, ed., 1974: 284.

13 Dumézil 1954: 52; Dagron 1974: 332-3, 336-7, and Cameron 1976: 231.

14 Hyginus, *Fabulae* 136 'Polyidus'. It should be noted that the universality of the Berlin and Kay scheme of development has now been challenged (see, e.g., Moss 1989); it is possible, however, that it is still relevant to the present discussion. The question requires further investigation.

15 Dumézil (1954: 55-7) treats green and blue as doublets and relates them both to the third function, but as, unlike Dumézil, I do not consider that the female is included in the third function,

I find that green has a different reference from blue.

16 Tertullian, *De Spectaculis* 9; Isidore of Seville, *Etymologia* 18.33; Malalas, ed. Dindorf 1831: 173-6; Lydus, *De Mensibus*, ed. Wuensch 1898: 1.12.

17 Cassiodorus, *Variae Epistolae* 3.51, trans. Hodgkin 1886: 227. For illustrations of the turning-post, see Humphrey 1986, e.g. pages 43-6, 181, 205, 222 and 236.

18 Tertullian, *De Spectaculis* 9, and Isidore of Seville, *Etymologia* 18.41; cf. Corippus, *In laudem Iustini Augusti minoris* 1.322-9, Cassiodorus, *Variae Epistolae* 3.51; Lydus, *De Mensibus*, ed. Wuensch 1898: 4.30, and the discussion of these passages in Cameron 1976: 336-8.

19 Lévi-Strauss 1968: 10, 105; cf. Maybury-Lewis 1979: 26-9, 46-8.

20 Vat. Gr. 1056 (= Cod. Rom. 20) f. 177 fr. ιη', *Catalogus Codicum Astrologorum Graecorum*, 5, 3.127-8, and Codex Ambrosianus 886: C 222 inf., f. 42 fr. ση', as quoted and, in the second case, emended in Wuilleumier 1927: 184-7.

21 Malalas, ed. Dindorf 1831: 173-4 (see p. 130); cf. Lydus, *De Mensibus*, ed. Wuensch 1898: 1.12.

### 5 The Dark Days and the Light Month

1 See Aveni 1980: 67-71, Ashbrook 1984: 200-9, and Ilyas 1984.

2 This statement does not apply to the polar regions but they do not seem to be immediately relevant to the present discussion.

3 This would mean that the months ceased to be tied to the observation of the lunar cycles, as in the case of the Egyptian civil calendar; see Parker 1950.

4 For discussion of periods of reversal, see Frazer 1913: 306-54, and Babcock 1978.

5 *Rigveda* 4.33.7, discussed in Frazer 1913: 324-5.

### 6 Archaic Calendar Structure Approached
### Through the Principle of Isomorphism

1 There can be 355 days in a lunar year but this number is less frequent; see Michels 1967: 11-2.

2 The 'Egyptian calendar' referred to throughout this chapter is the civil calendar. Information about it has been drawn from Foucart 1910, Parker 1950: 51-6, and Griffiths, ed., 1970: 291-308.

### 7 *Distinctive Features in Cosmic Structure*

1 Claude Lévi-Strauss quoting Roman Jakobson in his preface to Jakobson 1978: xii.

### 8 *The Design of the Celtic Year*

1 Danaher 1982: 217-42. The modern dates given are, of course, only approximations to the dates of the pagan festivals.
2 Rees and Rees 1961: 100-3, 122-4, and Byrne 1973: 168. There were two legendary divisions of Ireland into halves, an original one attributed to the sons of Partholon when the division line ran roughly E-W along Eiscir Riada, and a later one attributed to the Fir Bolg when the division line ran roughly NE-SW separating the pairs of provinces (Macalister, ed., 1938-56: 3.22-5, 4.5). It is the second of these that is related to the seasons in *The Death of King Dermot*.
3 O'Grady 1892: 1.80-1, 2.86; cf. Atkinson 1896: 174.
4 For the definition of the axes, see chapter 7.
5 Loth 1904: 127. See Pokorny 1959-69: 1.425-6 '*g̑hei-*' = winter, Old Irish '*gamuin*', 1.1174 '*u̯es-*' = spring, Old Irish '*errach*'; 1.905 '*sem-*' = summer, Old Irish '*sam*'.
6 In terms of the system of axes, the effect of this change would be to collapse a three-axis system which distinguished a B axis (± maximum moonlight, i.e. full moon versus crescents) from a C axis (± moonlight, i.e. light month versus interlunium) into a two-axis system which no longer distinguished between the B and C axes, the C axis having been brought into correspondence with the B axis.

### 9 *Polarity, Deixis, and Cosmological Space and Time*

1 I am very grateful to Ann Harleman Stewart for introducing me to the literature on deixis at the American Folklore Society conference in Minneapolis in 1982. Her book (Stewart 1976) has also been generally helpful to me in studying and articulating structural ideas.

### 10 *Cyclical Time as Two Types of Journey and Some Implications for Axes of Polarity, Contexts, and Levels*

1 *Rigveda* 1.185.1; *Aitareya Brāhmaṇa* 3.44; Sieg 1923: 2, 8, 10; Kuiper 1983: 75, 83; Witzel 1984: 230. For the return journey of the sun across the sky at night in an African context, see Jellicoe

1985: 41-2.

2 I would like to offer the suggestion that the two opposite ways of
defining 'movement to the right' could be connected with the
starting point selected. If the hand is in the top position and begins
a circling movement to the right, the action is clockwise and,
conversely, if the hand is in the bottom position and begins a
circling movement to the right, the action is anti-clockwise. Cf.
the 'top of the head' and 'base of the tree' contrast in this chapter.

## 11 The Place of the Hostile Twins in a
### Proposed Theogonic Structure

1 The Greek tradition is studied in West 1975.

2 See Zaehner 1955: 54-79, 419-28. The fullest story, which is told
in *De Deo* by Eznik of Kolb, can be found in the original
Armenian with a French translation in the edition of Louis Maries
and Ch. Mercier, 1959. Watts (1969: 129-30) gives a translation
of the French into English.

3 Malalas, ed. Dindorf 1831: 171-2. So far as I am aware, little use
has been made of this passage in treatments of the Roman twins.
Dagron does refer to it in his discussion of the Byzantine hippodrome (1974: 338-44) but considers Romulus and Remus and
Castor and Pollux to be equivalent twins, and altogether sees a
less complex dual system expressed in the circus symbolism than
I do (see chapter 4).

4 183 and 389 n. 13; see also Griffiths 1959: 40, 55.

5 The occurrence of five young gods (rather than the six generally
discussed in this chapter) can be related to the needs of a calendar
which had twelve months of thirty days, giving 360 days, and
leaving five full days over within the solar year of $365\frac{1}{4}$ days (see
chapter 6).

## 12 Whites and Reds: The Roman Circus
### and Alternate Succession

1 The polarity above/below is the A axis of the three-axis system.

2 This narrative occurs in the Bonn edition of Malalas (ed. Dindorf
1831) at pages 173-4 and in the translation by Jeffreys, Jeffreys
and Scott (1986) at pages 92-3. I am indebted to Dr R.C. McCail
for the translation given here. Alan Cameron (1976: 64-6) notes
the use by Malalas of earlier sources, and a reference to Charax
in connection with a mention of the contest between Pelops and

Oenomaus elsewhere in the chronicle (ed. Dindorf 1831: 81; trans. Jeffreys, Jeffreys and Scott 1986: 39) suggests that Charax of Pergamum may have been the source of Malalas's material on this contest. It will be evident that no attempt is being made here to relate the narrative to what actually happened historically in early Rome; this may well be beyond recall (cf. Poucet 1985).

### 13 The Chinese Trigrams and Archaic Cosmology

1 For details, see Lǐ Yǎn and Dù Shírán 1987: 3-11, and Needham 1954-84: 3.5-17.

2 On Yin-Yang and the *ch'i*, see Graham 1986: 70-4, 77, 84-92, and Schwartz 1985: 353.

3 See Tzey-yueh Tain 1974: 89-92, 144-55, referring to Tung Chung-shu, ed. Su Yü 1910: 10.35.14b, 12.53.10a, and 15.70.7b-8a. I am very grateful to Dr John Chinnery of the Department of Chinese at the University of Edinburgh who spoke of the three 'fishnet cords' (as the term *kang* can be rendered) in one of the first talks given to the Traditional Cosmology Society in 1984, and to his colleague, Dr A.W.E. Dolby, who has kindly discussed the content of this chapter and helped me to locate references.

4 In comparison with the more familiar five relationships – king and minister, father and son, husband and wife, senior and junior, and friend and friend (see Ames 1983: 225 n. 41, 247-8 n. 129) – these three could either be a subset, or, as I would suggest, the foundation set from which the more extended set was developed.

5 There is no need to import with these terms any greater positive and negative feeling than is present in a particular system. Graham notes (1986: 28) that the corresponding yang and yin items in the binary chain are 'mutually dependent' but that the yang item is 'superior to' the yin item.

6 For a study of the relationship between various moiety systems as found among the Eastern Timbara of Central Brazil, see Lave 1977. On possible moiety systems in the Shang context, see Chang 1978, and Wheatley 1971: 53-4, 98 n. 162, quoting Liu Pin-hsiung 1965: 106-8.

7 Keightley 1987; Li Xueqin 1985: 489. Schwartz (1985: 454, n. 5) comments in relation to the Shang: 'The concern with the "four cardinal directions" in the oracle bone inscriptions is taken by some to have some connection to correlative cosmology.'

8 Granet 1950: 170; *Shuo-wên chü-tou*, ed. Wang Yun 1983: 4.2122-64. Graham (1986: 49-50) comments in connection with the sets of five that 'the basic correlation is of the Four Seasons with the Four Directions' and discusses the proportional opposition involved.

9 Girschner 1912: 199-200; Bollig 1927: 65; Lessa 1959: 190.

10 In using 1 and 2 to indicate yang and yin lines, I am following the suggestion that the 'complete' line is basically a single stroke and that the 'broken' line is basically two strokes; see Vandermeersch 1974: 47. Other ways of representing yang and yin would not affect the argument, however.

## 14 Transformations

1 For discussion of the use of the analogy of the human body in Jewish cosmological theory, see Idel 1988: 112-22.

2 In making comparisons between analogical series of items, account should be taken of the debate on homologous terms and homologous relations; see Needham 1980: 41-62, and Racine 1989. In this discussion, I am making use of the fact that the same qualities are attributed to items in the different series.

3 Cf. Keith Hutchison's remarks about precedents for the Copernican concept of the central sun (Hutchison 1987: 104-5 and n. 16).

4 The view of horizontal space includes right and left, and it is noteworthy that G.E.R. Lloyd comments (1973: 174) on 'the faithfulness, one may almost say stubbornness, with which Aristotle adheres to his conception of the essential superiority of right to left' and adds (179):

It is, perhaps, particularly remarkable that Aristotle, who conducted the most extensive and rigorous investigations in antiquity, should nevertheless have firmly and constantly maintained a theory of the distinction between right and left which owes much to the traditional symbolic associations which those opposites had for the ancient Greeks.

# References

Akatsuko Kiyoshi (1982). On the Cosmological Meaning of the Calendar Signs 'Shi Gan' and 'Shi-er Zhi'. Typescript paper; for details of availability, consult the next item.

— (1986). The Cosmological Meaning of the ten gan and twelve zhi in Shang Civilization. Paper summary. *Early China*, Supplement 1, International Conference on Shang Civilization, 1982, pp. 38-40.

Alkire, William H. (1970). Systems of Measurement on Woleai Atoll, Caroline Islands. *Anthropos 65*, 1-73.

Allen, N.J. (1985). Hierarchical Opposition and Some Other Types of Relation. In R.H. Barnes and others, pp. 21-32.

— (1987). The Ideology of the Indo-Europeans: Dumézil's Theory and the Idea of a Fourth Function. *International Journal of Moral and Social Studies 2*, 23-39.

Ames, Roger T. (1983). *The Art of Rulership: A Study in Ancient Chinese Political Thought*. Honolulu: University of Hawaii Press.

Apollodorus, trans. Sir James George Frazer (1921). *The Library*. London and New York: William Heinemann and G.P. Putnam's Sons.

Aristotle, trans. H.D.P. Lee (1952). *Meteorologica*. London and Cambridge, Mass.: William Heinemann and Harvard University Press.

—, trans. A.L. Peck and E.S. Forster (1937). *Parts of Animals; Movement of Animals; Progression of Animals*. London and Cambridge, Mass.: William Heinemann and Harvard University Press.

—, trans. Ernest Barker (1946). *The Politics*. Oxford: Clarendon Press.

Ashbrook, Joseph (1984). *The Astronomical Scrapbook: Skywatchers, Pioneers, and Seekers in Astronomy*. Cambridge and Cambridge, Mass.: Cambridge University Press and Sky Publishing

Corporation.

Athanassakis, Apostolos N., trans. (1977). *The Orphic Hymns.* Missoula, Montana: Scholars Press.

Atkinson, R., ed. (1896). *The Yellow Book of Lecan.* Dublin: Royal Irish Academy.

Aveni, Anthony F. (1980). *Skywatchers of Ancient Mexico.* Austin: University of Texas Press.

Babcock, Barbara A. (1978). *The Reversible World.* Ithaca and London: Cornell University Press.

Bannerman, John (forthcoming). The *Clarsach* and the *Clarsair.* *Scottish Studies.*

Bannister, D. and Fay Fransella (1971). *Inquiring Man: The Theory of Personal Constructs.* Harmondsworth: Penguin.

Barnes, R.H. (1985). Hierarchy without Caste. In R.H. Barnes and others, pp. 8-20.

Barnes, R.H. and others (1985). *Contexts and Levels: Anthropological Essays on Hierarchy.* Oxford: JASO.

Basham, A.L. (1967). *The Wonder That Was India.* 3rd ed. London: Sidgwick and Jackson.

Beck, Brenda E.F.(1969). Colour and Heat in South Indian Ritual. *Man,* n.s. 4, 553-72.

van Beek, W.E.A. (1979). Traditional Religion as a Locus of Change. In *Official and Popular Religion,* ed. Pieter Hendrik Vrithof and Jacques Waardenburg (The Hague, Paris and New York: Mouton), pp. 514-43.

Benveniste, Émile (1932). Les classes sociales dans la tradition avestique. *Journal asiatique* 221, 117-34.

— (1938). Traditions indo-iraniènnes sur les classes sociales. *Journal asiatique* 230, 529-49.

— (1973). *Indo-European Language and Society.* London: Faber and Faber.

Berlin, Brent and Paul Kay (1969). *Basic Color Terms: Their Universality and Evolution.* Berkeley and Los Angeles: University of California Press.

Bernardi, Bernardo (1985). *Age Class Systems.* Cambridge: Cambridge University Press.

Bianchi, Ugo (1971). Seth, Osiris et l'ethnographie. *Revue de l'histoire des religions* 179, 113-35. Also in Bianchi (1978), 103-25.

— (1978). *Selected Essays on Gnosticism, Dualism and Mysteriosophy.* Leiden: E.J. Brill.

Binchy, D.A. (1958). The Fair of Tailtiu and the Feast of Tara. *Eriu* 18, 113-38.

Bjornsson, Arni (1980). *Icelandic Feasts and Holidays*. Reykjavik.

Bloch, Maurice (1979). Knowing the world or hiding it. *Man* n.s. 14, 165-7.

Bloch, Maurice and Jonathan Parry (1982). *Death and the Regeneration of Life*. Cambridge: Cambridge University Press.

Blom, Eric, ed. (1954). *Grove's Dictionary of Music and Musicians*. 5th ed. London: Macmillan and Co.

Boethius, Anicius Manlius Severinus, ed. Godofredus Friedlein (1966 [1867]). *De Institutione Arithmetica. De Institute Musica*. Frankfurt am Main: Minerva G.M.B.H.

Bollig, Laurentius (1927). *Die Bewohner der Truk-Inseln*. Anthropos Ethnologische Bibliothek, Vol. 3, Pt. 1. Münster: Aschendorff.

Bosch, F.D.K. (1960). *The Golden Germ: An Introduction to Indian Symbolism*. 's-Gravenhage: Mouton.

Boyce, Mary (1975). *A History of Zoroastrianism*, Vol. 1 The Early Period. Leiden/Köln: Brill, Handbuch der Orientalistik 1.8.1.2.2A.

Bronson, Bertrand H. (1959-72). *The Traditional Tunes of the Child Ballads*. Princeton, N.J.: Princeton University Press.

van Buitenen, J.A.B. (1968) *The Pravargya*. Poona: Deccan College Postgraduate and Research Institute.

Burkert, Walter (1962). *Weisheit und Wissenschaft: Studien zu Pythagoras, Philolaos und Platon*. Nuremberg: Erlanger.

— (1972). *Lore and Science in Ancient Pythagoreanism*. Cambridge, Mass.: Harvard University Press.

Burton, Anne (1972). *Diodorus Siculus Book 1: A Commentary*. Leiden: Brill.

Byrne, Francis (1973). *Irish Kings and High Kings*. London: Batsford.

Byrne, Mary E. and Myles Dillon, ed. (1937). *Táin Bó Fraích. Etudes Celtiques* 2, 1-27.

Caland, W. trans. (1969 [1928]). *Das Srautasūtra des Āpastamba: Sechzehntes bis Vierundzwanzigstes und Einunddreissigstes Buch*. Wiesbaden: Martin Sändig.

Caldwell, John (1981). The *De Institutione Arithmetica* and the *De Institutione Musica*. In *Boethius: His Life, Thought and Influence*, ed. Margaret Gibson (Oxford: Basil Blackwell), pp. 135-54.

Cameron, Alan (1976). *Circus Factions: Blues and Greens at Rome and Byzantium*. Oxford: Clarendon Press.

Cassiodorus, Magnus Aurelius Senator, ed. A.J. Fridh and J.W. Halporn (1973). *Opera*, Part 1, *Variarum Libri XII* and *De Anima*. Corpus Christianorum Series Latina 96. Turnholt: Brehols.

—, trans. Thomas Hodgkin (1886). *The Letters of Cassiodorus, being a condensed translation of the Variae Epistolae of Magnus Aurelius Cassiodorus Senator*. London: Henry Frowde.

Cassirer, Ernst (1955). *The Philosophy of Symbolic Forms*, Vol. 2, Mythical Thought. New Haven and London: Yale University Press.

Cedrenus, George, ed. Immanuel Bekker (1838-9). *Compendium Historiarum*. Bonn: Weber, Corpus Scriptorum Historiae Byzantinae.

Chadwick, Henry (1981). *Boethius: the consolations of music, logic, theology and philosophy*. Oxford: Clarendon Press.

Chang, Kwang-chih (1976). Some Dualistic Phenomena in Shang Society. In Kwang-chih Chang, *Early Chinese Civilization: Anthropological Perspectives* (Cambridge, Mass. and London: Harvard University Press), pp. 93-114.

— (1978). *T'ien kan*: a key to the history of the Shang. In *Ancient China: Studies in Early Civilization*, ed. David T. Roy and Tsuenhsuin Tsien (Hong Kong: Chinese University Press), pp. 13-42.

— (1980). *Shang Civilization*. New Haven and London: Yale University Press.

Chao, Lin (1982). *The Socio-Political Systems of the Shang Dynasty*. Nankang, Taipei, Taiwan: Institute of the Three Principles of the People, Academia Sinica, Monograph Series No. 3.

Chelhod, J. (1973). A Contribution to the Problem of the Pre-eminence of the Right, Based upon Arabic Evidence. In Needham, ed., pp. 239-62.

Child, Francis James (1882-98). *The English and Scottish Popular Ballads*. Boston: Houghton, Mifflin and Co.

Choain, Jean (1983). *Introduction au Yi King*. Monaco: Editions du Rocher.

Claessen, Henri J.M. and Peter Skalnik, ed. (1978). *The Early State*. The Hague, Paris and New York: Mouton.

Clark, Eve V. (1979). *The Ontogenesis of Meaning*. Wiesbaden: Akademische Verlagsgesellschaft.

Clark, Herbert H. (1973). Space, Time, Semantics, and the Child. In *Cognitive Development and the Acquisition of Language*, ed.

Timothy E. Moore (New York and London: Academic Press), pp. 27-63.

Clark, Herbert H. and Eve V. Clark (1977). *Psychology and Language: An Introduction to Psycholinguistics*. New York: Harcourt Brace Jonanovich.

Colson, F.H. (1926). *The Week*. Cambridge: Cambridge University Press.

Corippus, Flavius Cresconius, ed. and trans. Averil Cameron (1976). *In laudem Iustini Augusti minoris*. London: Athlone Press.

Cornford, Francis M. (1937). *Plato's Cosmology: The 'Timaeus' of Plato translated with a running commentary*. London: K. Paul, Trubner and Co.

Dagron, Gilbert (1974). *Naissance d'une capitale: Constantinople et ses institutions de 330 à 451*. Paris: Presses universitaires de France.

Danaher, Kevin (1982). Irish Folk Tradition and the Celtic Calendar. In *The Celtic Consciousness*, ed. Robert O'Driscoll (Portlaoise: Dolmen Press), pp. 217-42.

Dicks, D.R. (1970). *Early Greek Astronomy to Aristotle*. Ithaca, New York: Cornell University Press.

Dijksterhuis, E.J. (1961). *The Mechanization of the World*. Oxford: Clarendon Press.

Dilke, O.A.W. (1971). *The Roman Land Surveyors: An Introduction to the 'Agrimensores'*. Newton Abbot: David and Charles.

Dillon, Myles (1975). *Celts and Aryans: Survivals of Indo-European Speech and Society*. Simla: Indian Institute of Advanced Study.

Diodorus of Sicily, trans. C.H. Oldfather (1933). *The Library of History*. Vol. 1. London and New York: William Heinemann and G.P. Putnam's Sons.

Dubuisson, Daniel (1985). Matériaux pour une typologie des structures trifonctionelles. *L'Homme* 93, 105-21.

Duff-Cooper, Andrew (1985). Ethnographical Notes on Two Operations of the Body Among a Community of Balinese on Lombok. *Journal of the Anthropological Society of Oxford* 16, 121-42.

Dumézil, Georges (1930). La préhistoire indo-iranienne des castes. *Journal asiatique* 216, 109-30.

— (1941). *Jupiter, Mars, Quirinus: essai sur la conception indo-européene de la société et sur les origines de Rome*. Paris: Gallimard.

— (1947). *Tarpeia: cinq essais de philologie comparative indo-*

*européenne*. Paris: Gallimard.

— (1948). *Mitra-Varuṇa: essai sur deux représentations indo-européennes de la souveraineté*. 2nd ed. Paris: Gallimard.

— (1954). *Rituels indo-européens à Rome*. Paris: Klincksieck.

— (1958). *L'idéologie tripartie des Indo-Européens*. Collection Latomus, vol. 31. Brussels: Latomus.

— (1969). *Idées romaines*. Paris: Gallimard.

— (1970). *Archaic Roman Religion*. Chicago and London: University of Chicago Press.

— (1973a). *The Destiny of a King*. Chicago and London: University of Chicago Press.

— (1973b). *Gods of the Ancient Northmen*. Berkeley, Los Angeles, and London: University of California Press.

— (1977). *Les dieux souverains des Indo-Européens*. Paris: Gallimard.

Dumont, Louis (1971). On Putative Hierarchy and Some Allergies To It. *Contributions to Indian Sociology* n.s. 5, 58-81.

— (1979). The anthropological community and ideology. *Social Science Information* 18, 83-110.

Dumont, P.-E. (1927). *L'Aśvamedha: Description de sacrifice solennel du cheval dans le culte Védique*. Paris: P. Guenther.

Durkheim, Emile and Marcel Mauss, trans. and ed. Rodney Needham (1963). *Primitive Classification*. London: Cohen and West.

Dyson-Hudson, Neville (1963). The Karimojong Age System. *Ethnology* 2, 353-401.

Eliade, Mircea (1955). *The Myth of the Eternal Return*. London: Routledge and Kegan Paul.

— (1978-86). *A History of Religious Ideas*. Chicago: University of Chicago Press.

Ellen, Roy (1986). Macrocosm, Microcosm and the Nuaula House: Concerning the Reductionist Fallacy as Applied to Metaphorical Levels. *Bijdragen tot de Taal-, Land- en Volkenkunde* 142, 1-30.

Eznik of Kolb (1959). *De Deo*, ed. Louis Maries and Ch. Mercier. Paris: Patrologia Orientalis, vol. 28, parts 3-4.

Farmer, H.G. (1925-6). The Influence of Music: From Arabic Sources. *Proceedings of the Musical Association* 52, 89-124.

— (1933). *An Old Moorish Lute Tutor*. Glasgow: Civic Press.

Filliozat, Jean (1953). Médecine. In *L'Inde classique* by Louis Renou and Jean Filliozat, vol 2 (Paris and Hanoi: Imprimerie Nationale and Ecole Française d'Extrême-Orient), pp. 139-66.

— (1964). *The Classical Doctrine of Indian Medicine: Its Origins and its Greek Parallels*. Delhi: Munshiram Manoharlal.

— (1975). *La doctrine classique de la médecine indienne; ses origines et ses paralleles grecs*. 2nd ed. Paris: Ecole Française d'Extrême-Orient.

Fillmore, Charles J. (1972). How to Know Whether You Are Coming or Going. In *Linguistik 1971*, ed. Karl Hyldgaard-Jensen (Frankfurt: Athenaum), pp. 369-79.

Forsberg, Nils (1943). *Une forme élémentaire d'organisation cérémoniale*. Uppsala: Almqvist and Wiksell.

Forth, Gregory L. (1981). *Rindi: An Ethnographic Study of a Traditional Domain in Eastern Sumba*. The Hague: Martinus Nijhoff.

— (1983). Time and Temporal Classification in Rindi, Eastern Sumba. *Bijdragen tot de Taal-, Land- en Volkenkunde* 139, 46-80.

— (1985). Right and Left as a Hierarchical Opposition: Reflections on Eastern Sumbanese Hairstyles. In R.H. Barnes and others, pp. 103-16.

Foucart, George (1910). Calendar (Egyptian). In *The Encyclopaedia of Religion and Ethics*. James Hastings (ed.), 3.91-105. Edinburgh: T. and T. Clark.

Fox, Denton, ed. (1981). *The Poems of Robert Henryson*. Oxford: Clarendon Press.

Frame, Douglas (1978). *The Myth of Return in Early Greek Epic*. New Haven and London: Yale University Press.

Frankfort, Henri (1948). *Kingship and the Gods*. Chicago: University of Chicago Press.

Frazer, Sir James George (1911). *The Golden Bough*. 3rd ed. Part III: *The Dying God*. London: Macmillan.

— (1913). *The Golden Bough*. 3rd ed. Part VI: *The Scapegoat*. London: Macmillan.

Fung, Yu-lan (1952-3). *A History of Chinese Philosophy*. Princeton: Princeton University Press.

Ghirshman, Roman (1962). *Iran: Parthians and Sassanians*. London: Thames and Hudson.

Ginzberg, Louis (1909-38). *The Legends of the Jews*. Philadelphia: Jewish Publication Society of America.

Ginzel, F.K. (1906-14). *Handbuch der Mathematischen und Technischen Chronologie*. Leipzig: J.C. Hinrichs.

Girschner, Max (1912). Die Karolineninsel Námōluk und ihre Bewohner. *Baessler-Archiv* 2, 123-215.

Givón, Talmy (1973). The Time-Axis Phenomenon. *Language* 49, 890-925.

Gonda, J. (1966). *Ancient Indian Kingship from the Religious Point of View.* Leiden: Brill.

— (1976). *Triads in the Veda.* Amsterdam, Oxford, and New York: North Holland.

Gosvami, O. (1957). *The Story of Indian Music: Its Growth and Synthesis.* Bombay: Asia Publishing House.

Graham, A.C. (1986). *Yin-Yang and the Nature of Correlative Thinking.* Singapore: Institute of East Asian Philosophies, National University of Singapore.

Granet, Marcel (1950). *La pensée chinoise.* Paris: Albin Michel.

— (1973). Right and Left in China. In Needham, ed., pp. 43-58.

Griffiths, J. Gwyn (1959). Some Remarks on the Enneads of Gods. *Orientalia* n.s. 28, 34-56.

— (1960). *The Conflict of Horus and Seth.* Liverpool: Liverpool University Press.

—, ed. (1970). *Plutarch's 'De Iside et Osiride'.* Cardiff: University of Wales Press.

Groot, J.J.M. de (1901). *The Religious System of China.* Volume 4. Leyden: E.J. Brill.

Gulliver, P.H. (1958). The Turkana Age Organization. *American Anthropologist* 60, 900-22.

Guthrie, W.K.C. (1950). *The Greeks and Their Gods.* London: Methuen and Co.

Hanfmann, G.M.A. (1951). *The Season Sarcophagus in Dumbarton Oaks.* Cambridge, Mass.: Harvard University Press.

Harris, H.A. (1972). *Sport in Greece and Rome.* London: Thames and Hudson.

— (1976) *Greek Athletics and the Jews.* Cardiff: University of Wales Press.

Harrison, Jane Ellen (1927). *Themis: A Study of the Social Origins of Greek Religion.* 2nd ed. Cambridge: Cambridge University Press.

Hastings, James, ed. (1908-26). *The Encyclopaedia of Religion and Ethics.* Edinburgh: T. and T. Clark.

Hayman, Peter (1986). *Sefer Yetsira (The Book of Creation). Shadow* 3, 20-38.

Held, Gerrit Jan (1935). *The Mahābhārata. An Ethnological Study.* Amsterdam: Uitgevers-maatschappij Holland.

Hesiod, trans. M.L. West (1988). *'Theogony' and 'Works and Days'.* Oxford and New York: Oxford University Press.

de Heusch, Luc (1981). *Why Marry Her?: Society and symbolic structures.* Cambridge: Cambridge University Press.

— (1985). *Sacrifice in Africa: A structuralist approach.* Manchester: Manchester University Press.

Homer, trans. A.T. Murray (1924-5). *The Iliad.* London and New York: William Heinemann and G.P. Putnam's Sons.

Howe, Leopold E.A. (1981). The Social Determination of Knowledge: Maurice Bloch and Balinese Time. *Man* n.s. 16, 220-34.

Hull, Vernam (1949). *Echtra Cormaic Maic Airt*, 'The Adventure of Cormac Mac Airt'. *Publications of the Modern Language Association* 64, 871-83.

Humphrey, John H. (1986). *Roman Circuses: Arenas for Chariot Racing.* London: B.T. Batsford.

Hutchison, Keith (1987). Towards a Political Iconology of the Copernican Revolution. In *Astrology, Science and Society,* ed. Patrick Curry (Boydell Press, Woodbridge, Suffolk), pp. 95-141.

Hyginus, ed. H.I. Rose (1963). *Fabulae.* Leyden: A.W. Sythoff.

Idel, Moshe (1988). *Kabbalah: New Perspectives.* New Haven and London: Yale University Press.

Ilyas, Mohammad (1984). *A Modern Guide to Astronomical Calculations of Islamic Calendar, Times and Qibla.* Kuala Lumpur: Berita.

Isidore of Seville, ed. W.M. Lindsay (1911). *Isidori Hispalensis episcopi etymologiarum sive originum libri XX.* Oxford: Clarendon Press.

Jakobson, Roman (1978). *Six Lectures in Sound and Meaning.* Hassocks, Sussex: Harvester Press.

Jakobson, Roman, with E. Colin Cherry and Morris Halle (1971). Toward the Logical Description of Languages in their Phonemic Aspect. In Roman Jakobson, *Selected Writings. I Phonological Studies.* 2nd ed. The Hague and Paris: Mouton.

Jellicoe, Marguerite (1985). Colour and Cosmology among the Nyaturu of Tanzania. *Shadow* 2, 37-44.

Jessen, Marilyn E. (1975). A Semantic Study of Spatial and Temporal Expressions in English. Ph.D. thesis, University of Edinburgh.

Jung, C.G. (1916). *Psychology of the Unconscious.* New York:

Moffat, Yard and Co.

Keightley, David N. (1978). *Sources of Shang History*. Berkeley, Los Angeles, and London: University of California Press.

— (1987). Astrology and Cosmology in the Shang Oracle-Bone Inscriptions. *Cosmos* 3, 36-40.

Keith, A. Berriedale (1924). *The Sanskrit Drama*. Oxford: Clarendon Press.

Kolinski, M. (1974). Modes, Musical. In the *Encyclopaedia Britannica*. 15th ed. (Chicago: Benton), *Macropaedia* 12.295-8.

Kuiper, F.B.J. (1961). Some Observations on Dumézil's Theory (with reference to Prof. Frye's article). *Numen* 8, 34-45.

—, ed. John Irwin (1983). *Ancient Indian Cosmogony*. Delhi: Vikas Publishing House.

Lave, Jean (1977). Eastern Timbara Moiety Systems in Time and Space: A Complex Structure. In *Actes du XLIIe Congrès International des Américanistes*, Vol. 2 *Social Time and Social Space in Lowland Southamerican Societies*, organised by Joanna Overing Kaplan (Paris: Société des Américanistes, Musée de l'Homme), pp. 309-21.

Leach, E.R. (1961). *Rethinking Anthropology*. London: Athlone Press.

Leach, Maria and Jerome Fried (1950). *Funk and Wagnalls Standard Dictionary of Folklore, Mythology and Legend*. New York: Funk and Wagnalls Company.

— (1975). *Funk and Wagnalls Standard Dictionary of Folklore, Mythology and Legend*. One-volume ed. London: New English Library.

Legge, J. trans., ed. Ch'u Chai and W. Chai (1969). *I Ching*. New York: University Books, Bantam Paperbacks.

Lessa, William A. (1959). Divining from Knots in the Carolines. *Journal of the Polynesian Society* 68, 188-205.

— (1969). The Chinese Trigrams in Micronesia. *Journal of American Folklore* 82, 353-62.

Lévi-Strauss, Claude (1968). *Structural Anthropology*. New York and London: Basic Books.

Li Xueqin, trans. K.C. Chang (1985). *Eastern Zhou and Qin Civilizations*. New Haven and London: Yale University Press.

Lǐ Yǎn and Dù Shírán, trans John N. Crossley and Anthony W.-C. Lun (1987). *Chinese Mathematics: A concise history*. Clarendon Press: Oxford.

Lincoln, Bruce (1975-6). The Indo-European Myth of Creation. *History of Religions* 15, 121-45.

— (1986). *Myth, Cosmos, and Society: Indo-European Themes of Creation and Destruction*. Cambridge, Mass. and London: Harvard University Press.

Littlejohn, James (1973). Temne Right and Left: An Essay on the Choreography of Everyday Life. In Needham 1973, pp. 288-98.

Littleton, C. Scott (1982). *The New Comparative Mythology: An Anthropological Assessment of the Theories of Georges Dumézil*. 3rd ed. Berkeley, Los Angeles, and London: University of California Press.

Liu Pin-hsiung (1965). Yin-Shang Wang-shih shih-fen-tsu-chih shih-lun. *Chung-yang Yen-chiu-yüan: Min-ts'u-hsüeh Yen-chiu-so Chi-k'an* 19, 89-114.

Livy, trans. B.O. Foster (1919). *Livy*, Books 1-2. London and New York: William Heinemann.

Lloyd, G.E.R. (1966). *Polarity and Analogy: Two Types of Argumentation in Early Greek Thought*. Cambridge: Cambridge University Press.

— (1973). Right and Left in Greek Philosophy. In Needham, ed., pp. 167-86.

—, ed. (1983). *Hippocratic Writings*. Harmondsworth: Penguin.

Loth, J. (1904). L'année celtique. *Revue celtique* 25, 113-62.

Lotman, J.M. (1968). Sémantique du nombre et type de culture. *Tel quel* 35, 24-7.

Lydus, John, ed. R. Wuensch (1898). *Liber de Mensibus*. Leipzig: B.G. Teubner.

Lyle, Emily (1979). Cosmos and Indo-European Folktales. In *Abstracts of Papers. Seventh Congress of the International Society for Folk-Narrative Research, Edinburgh, Scotland, 12th-18th August*, p. 78.

Lyons, John (1977). *Semantics*. Cambridge: Cambridge University Press.

Macalister, R.A.S., ed. (1938-56). *Lebor Gabála Erenn*. Dublin: Irish Texts Society.

Mac Cana, Proinsias (1983). *Celtic Mythology*. 2nd ed. Feltham, Middlesex: Newnes.

MacNeill, Eoin, ed. (1908). *Duanaire Finn: The Book of the Lays of Fionn*, vol. 1. London: Irish Texts Society.

MacQueen, John (1976). Neoplatonism and Orphism in Fifteenth-

Century Scotland. The Evidence of Henryson's *New Orpheus.* *Scottish Studies* 20, 69-89.

*Mahābhārata, The,* trans. and ed. J.A.B. van Buitenen (1973-). Chicago: University of Chicago Press.

Major, John (1984). The Five Phases, Magic Squares, and Schematic Cosmography. In *Explorations in Early Chinese Cosmology,* ed. Henry Rosemont, Jr. (Chico, California: Scholars Press), pp. 133-66.

Malalas, John, ed. Ludovic Dindorf (1831). *Chronographia.* Bonn: Weber, Corpus Scriptorum Historiae Byzantinae.

—, trans. Elizabeth Jeffreys, Michael Jeffreys and Roger Scott (1986). *The Chronicle of John Malalas.* Melbourne: Australian Association for Byzantine Studies, Byzantina Australiensia 4.

Malamoud, Charles (1981). On the rhetoric and semantics of puruṣārtha. *Contributions to Indian Sociology* n.s. 15, 33-54.

Maybury-Lewis, David H.P. , ed. (1979). *Dialectical Societies: The Ge and Bororo of Central Brazil.* Cambridge, Mass. and London: Harvard University Press.

— (1985). On theories of order and justice in the development of civilization. *Symbols: a publication of the Peabody Museum and the Department of Anthropology, Harvard University,* December 1985, pp. 17-21.

Maybury-Lewis, David H.P. and Uri Almagor, ed. (1989). *The Attraction of Opposites: Thought and Society in a Dualistic Mode.* Ann Arbor: University of Michigan Press.

Meid, Wolfgang, ed. (1970). *Die Romanze von Froech und Findabair: Táin Bó Froích.* Innsbruck: Gesellschaft zur Pflege der Geisteswissenschaften.

Mercer, Samuel A.B. (1952). *The Pyramid Texts.* New York, London, Toronto: Longmans, Green and Co.

Meyer-Baer, Kathi (1970). *Music of the Spheres and the Dance of Death: Studies in Musical Iconology.* Princeton, N.J.: Princeton University Press.

Michels, Agnes Kirsopp (1967). *The Calendar of the Roman Republic.* Princeton, N.J.: Princeton University Press.

Miller, George A. and Philip N. Johnson-Laird (1976). *Language and Perception.* Cambridge, Mass.: Belknap Press, Harvard University Press.

Moss, A.E. (1989). Basic Colour Terms: Problems and Hypotheses. *Lingua* 78, 313-20.

Müller, Werner (1961). *Die Heilige Stadt*. Stuttgart: Kohlhammer.

Muni, Bhārata, trans. Manomohan Ghosh (1950-61). *The Nātyaśāstra*. Calcutta: Royal Asiatic Society of Bengal.

Müri, Walter (1953). Melancholie und schwarze Galle. *Museum Helveticum* 10, 21-38. Reprinted in *Antike Medizin*, ed. Hellmut Flashar (Darmstadt: Wissenschaftliche Buchgesellschaft, 1971), pp. 165-91.

Murphy, Gerard (1953). *Duanaire Finn: The Book of the Lays of Fionn*, vol. 3. Dublin: Irish Texts Society.

Murphy, N.R. (1951). *The Interpretation of Plato's 'Republic'*. Oxford: Clarendon Press.

Nagy, Gregory (1980). Patroklos, Concepts of Afterlife, and the Indic Triple Fire. *Arethusa* 13, 161-95.

— (1986). Sovereignty, Boiling Cauldrons, and Chariot-Racing in Pindar's *Olympian* 1. *Cosmos* 2, 143-7.

Needham, Joseph, ed. (1954-84). *Science and Civilisation in China*. Cambridge: Cambridge University Press.

Needham, Rodney, ed. (1973). *Right & Left: Essays on Dual Symbolic Classification*. Chicago and London: Chicago University Press.

— (1980). *Reconnaissances*. Toronto and London: University of Toronto Press.

— (1983). Alternation. In Rodney Needham, *Against the 'Tranquility' of Axioms* (Berkeley, Los Angeles, and London: University of California Press), pp. 121-54.

— (1985). Dumézil and the Scope of Comparativism. In Rodney Needham, *Exemplars* (Berkeley, Los Angeles, and London: University of California Press), pp. 178-87.

Netting, Robert McC. (1972). Sacred Power and Centralization: Aspects of Political Adaptation in Africa. In *Population Growth: Anthropological Implications*, ed. Brian Spooner (Cambridge, Mass. and London: MIT Press), pp. 219-44.

Nicolas, Guy (1966). Essai sur les structures fondamentales de l'espace dans la cosmologie Hausa. *Journal de la Société des Africanistes* 36, 65-108.

— (1968) Un système numérique symbolique: le quatre, le trois et le sept dans la cosmologie d'une société hausa (vallée de Maradi). *Cahiers d'études africaines* 8, no. 32, 566-616.

Nilsson, Martin P. (1920). *Primitive Time-Reckoning*. Lund and London: C.W.K. Gleerup and Humphrey Milford.

Norberg-Schulz, Christian (1971). *Existence, Space and Architecture*. London: Studio Vista.

O'Curry, Eugene (1873). *On the Manners and Customs of the Ancient Irish*. London, Edinburgh, and Dublin: Williams and Norgate.

O Daly, Mairin, ed. (1975). *Cath Maige Mucrama: The Battle of Mag Mucrama*. Dublin: Irish Texts Society.

O'Flaherty, Wendy Doniger (1975). *Hindu Myths*. Harmondsworth: Penguin Books.

— (1979-80). Sacred Cows and Profane Mares in Indian Mythology. *History of Religions* 19, 1-25.

O'Grady, Standish H. (1892). *Silva Gadelica*. London and Edinburgh: Williams and Norgate.

Osler, Margaret J. and J. Brookes Spencer (1985). History of the physical sciences. In *The New Encyclopaedia Britannica*, 15th ed. (Encyclopaedia Britannica, Inc., Chicago), 25.834-41.

Parker, Richard A. (1950). *The Calendars of Ancient Egypt*. Chicago: University of Chicago Press.

Philo of Alexandria, trans. F.H. Colson and G.H. Whitaker (1929-62). *Philo*. London and New York / Cambridge, Mass.: William Heinemann and G.P. Putnam's Sons / Harvard University Press.

Plato, trans. Walter Hamilton (1973). *Phaedrus*. Harmondsworth: Penguin.

—, ed. John Ferguson (1957). *Republic Book X*. London: Methuen.

—, trans. Paul Shorey (1930-5). *The Republic*. London and Cambridge, Mass.: William Heinemann and Harvard University Press.

—, trans. H.D.P. Lee (1955). *The Republic*. Harmondsworth: Penguin.

—, trans. B. Jowett (1888). *The Republic*. 3rd ed. Oxford: Clarendon Press.

Plutarch, ed. F.C. Babbitt and others. *Moralia*. Cambridge, Mass. and London: Harvard University Press.

Pokorny, Julius (1959-69). *Indogermanisches Etymologisches Wörterbuch*. Berne and Munich: Francke.

Poucet, Jacques (1985). *Les origines de Rome: Tradition et histoire*. Brussels: Facultés universitaires Saint-Louis.

Puhvel, Jaan, ed. (1970). *Myth and Law among the Indo-Europeans: Studies in Indo-European Comparative Mythology*. Berkeley, Los Angeles, and London: California University Press.

— (1970). Aspects of Equine Functionality. In Puhvel 1970, pp. 159-

72.
— (1975-6). *Remus et frater*. *History of Religions* 15, 146-57.
— (1987). *Comparative Mythology*. Baltimore and London: Johns Hopkins University Press.
Quandt, Guilelmus (1962). *Orphei Hymni*. Berlin: Weidmann.
Quinn-Schofield, W.K. (1968). The Metae of the Circus Maximus as a Homeric Landmark – *Iliad*, 23.3²⁷-333. *Latomus* 27, 142-6.
Racine, Luc (1989). Du Modèle analogique dans l'analyse des représentations magico-religieuses. *L'Homme* 109, 5-25.
Radhakrishnan, S. (1940). *Eastern Religions and Western Thought*. 2nd ed. London: Oxford University Press.
Rees, Alwyn and Brinley Rees (1961). *Celtic Heritage: Ancient Tradition in Ireland and Wales*. London: Thames and Hudson.
Reichard, Gladys A. (1950). *Navaho Religion*. New York: Pantheon Books.
Richardson, N.J., ed. (1974). *The Homeric Hymn to Demeter*. Oxford: Clarendon Press.
Riese, Alexander (1869). *Anthologia Latina*. Leipzig: B.G. Teubner.
*Rigveda, The*, trans. Adolf Kaegi (1886). Boston, Mass.: Ginn.
Rimmer, Joan (1969). *The Irish Harp*. Cork: The Mercier Press.
Roscher, W.H. (1903). Die Enneadischen und Hebdomadischen Fristen und Wochen der Ältesten Griechen. *Der Abhandlungen der philologisch-historischen Klasse der Königl. Sächsischen Gesellschaft der Wissenschaften* 21 (4). Leipzig: B.G. Teubner.
Rowe, Christopher (1972). Concepts of Colour and Colour Symbolism in the Ancient World. *Eranos Yearbook* 41, 327-64.
Ruud, Jørgen (1960). *Taboo: A Study of Malagasy Customs and Beliefs*. Oslo and London: Oslo University Press and George Allen and Unwin.
Sahlins, Marshall (1981). The Stranger-King or Dumézil among the Fijians. *Journal of Pacific History* 16, 107-32.
— (1985). *Islands of History*. Chicago and London: University of Chicago Press.
de Santillana, Giorgio and Hertha von Dechend (1969). *Hamlet's Mill: An Essay on Myth and the Frame of Time*. Boston: Gambit.
Saso, Michael R. (1972). *Taoism and the Rite of Cosmic Renewal*. Purdue: Washington State University Press.
Sastri, H. Krishna (1974 [1916]). *South-Indian Images of Gods and Goddesses*. Madras: Government Press.
*Śatapatha Brāhmaṇa, The*, trans. Julius Eggeling (1882-1900). Sacred

Books of the East 12, 26, 41, 43, 44. Oxford: Clarendon Press.
de Saussure, Leopold (1909). *Les Origines de l'Astronomie Chinoise.* Leiden: Brill.
Schöner, Erich (1964). *Das Viererschema in der Antiken Humoralpathologie.* Wiesbaden: Franz Steiner Verlag GMBH.
Schrader, O. and A. Nehring (1917-29). *Reallexikon der Indogermanischen Altertumskunde.* 2nd ed. Berlin and Leipzig: Walter de Gruyter.
Schwartz, Benjamin I. (1985). *The World of Thought in Ancient China.* Cambridge, Massachusetts, and London: The Belknap Press of Harvard University Press:
Sergent, Bernard (1976). La représentation spartiate de la royauté. *Revue de l'histoire des religions* 189, 3-52.
de Shane, W. Mac Guckin, trans. (1843-71). *Ibn Khallikan's Biographical Dictionary.* Paris: Académie des Inscriptions et Belles-Lettres.
Shuldham-Shaw, Patrick (1976). The Ballad *King Orfeo. Scottish Studies* 20, 124-6.
Sieg, Emil (1923). Der Nachtweg der Sonne nach der vedischen Anschauung. *Nachrichten von der Königlichen Gesellschaft der Wissenschaften zu Göttingen, Philologisch-Historische Klasse 1923,* 1.1-23.
Speyer, J.S. (1906). A Remarkable Vedic Theory about Sunrise and Sunset. *Journal of the Royal Asiatic Society of Great Britain and Ireland,* pp. 723-7.
Stewart, Ann Harleman (1976). *Graphic Representation of Models in Linguistic Theory.* Bloomington: Indiana University Press.
Stewart, Marion (1973). *King Orphius. Scottish Studies* 17, 1-16.
Stokes, Whitley (1891-7). *The Second Battle of Moytura. Revue celtique* 12, 52-130.
Stokes, Whitley and E. Windisch, ed. (1891-7). *Irische Texte.* 3rd ser. Leipzig: S. Hirzel.
Sturluson, Snorri, trans. Samuel Laing, revised Peter Foote (1964). *Heimskringla: Part 2. Sagas of the Norse Kings.* London: Dent.
Syrkine, A.I. and V.N. Toporov (1968). La triade et la tétrade. *Tel quel* 35, 27-32.
Tacitus, trans. William Peterson and Maurice Hutton (1914). *Dialogus, Agricola, Germania.* London and New York: William Heinemann and Macmillan.
Terray, E. (1977). Event, structure and history: the formation of the

Abron kingdom of Gyaman (1700-1780). In *The Evolution of Social Systems*, ed. J. Friedman and M.J. Rowlands (London: Duckworth), pp. 279-301.

Tertullian, *'Apology' and 'De Spectaculis'* (1931), trans. T.R. Glover. London and New York: William Heinemann and G.P. Putnam's Sons.

Toporov, V.N. (1976). On the Cosmological Origins of Early Historical Descriptions. In *Russian Poetics in Translation*, Vol. 3 *General Semiotics*, ed. L.M. O'Toole and Ann Shukman (Oxford: Holdan Books), pp. 38-81.

Tracy, Theodore James (1969). *Physiological Theory and the Doctrine of the Mean in Plato and Aristotle*. The Hague: Mouton.

Traugott, Elizabeth Closs (1975). Spatial Expressions of Tense and Temporal Sequencing: A Contribution to the Study of Semantic Fields. *Semiotica* 15, 207-30.

Traugott, Elizabeth Closs and Mary Louise Pratt (1980). *Linguistics for Students of Literature*. New York: Harcourt Brace Jonanovich.

Trevarthen, Colwyn (1984). Dual Experience with a Two-Part Brain. *Shadow* 1, 17-21.

Tuan, Yi-Fu (1978). Space, Time, Place: A Humanistic Frame. In *Timing Space and Spacing Time*, Volume 1 *Making Sense of Time*, ed. Tommy Carlstein and others (London, Arnold), pp. 7-16.

Tung Chung-shu, ed. Su Yü (1910). *Ch'un-ch'iu fan-lu.*

Turner, Terence (1984). Dual Opposition, Hierarchy, and Value: Moiety Structure and Symbolic Polarity in Central Brazil and Elsewhere. In *Différences, valeurs, hiérarchie: textes offerts à Louis Dumont*, ed. Jean-Claude Galey (Paris: Ecole des Hautes Etudes en Sciences Sociales), pp. 335-70.

Turner, Victor (1967). Color Classification in Ndembu Ritual: A Problem in Primitive Classification. In Victor Turner, *The Forest of Symbols: Aspects of Ndembu Ritual* (Ithaca, New York: Cornell University Press), pp. 59-92.

Tzey-yueh Tain (1974). Tung Chung-shu's system of thought, its sources and its influence on Han scholars. Ph.D. thesis, University of California at Los Angeles. Ann Arbor, Michigan: University Microfilms.

Urwick, Edward J. (1920). *The Message of Plato: A Re-Interpretation of the 'Republic'*. London: Methuen and Co.

Vandermeersch, Léon (1974). De la tortue à l'achillée. In *Divination et rationalité*, ed. J.P. Vernant et al. (Paris: Editions du Seuil), pp.

29-51.

Vaughan, James H. (1980). A Reconsideration of Divine Kingship. In *Explorations in African Systems of Thought*, ed. Ivan Karp and Charles S. Bird (Bloomington: Indiana University Press), pp. 120-42.

te Velde, H. (1967). *Seth, God of Confusion*. Leiden: Brill.

Victor, Sextus Aurelius, ed. F. Pichlmayr, rev. R. Gruendel (1961). *Liber de Caesaribus; praecedunt, Origo gentis romanae et Liber de viris illustribus urbis Romae, subsequitur Epitome de Caesaribus*. Leipzig: B.G. Teubner.

Wang Yun, ed. (1983). *Shuo-wên chü-tou*. Shanghai: Ku-chi Shutien.

Ward, Donald (1968). *The Divine Twins: An Indo-European Myth in Germanic Tradition*. Berkeley and Los Angeles: University of California Press.

— (1970). The Threefold Death: An Indo-European Trifunctional Sacrifice? In Puhvel 1970, pp. 123-42.

Watts, Alan W. (1969). *The Two Hands of God: The Myths of Polarity*. New York: Collier Books.

Weiner, James F. (1988). *The Heart of the Pearl Shell: The Mythological Dimension of Foi Sociality*. Berkeley, Los Angeles, and London: University of California Press.

Wellesz, Egon, ed. (1957). *The New Oxford History of Music*, vol. 1 *Ancient and Oriental Music*. London: Oxford University Press.

West, M.L. (1971). *Early Greek Philosophy and the Orient*. Oxford: Clarendon Press.

— (1975). *Immortal Helen: An Inaugural Lecture*. London: Bedford College, University of London.

Wheatley, Paul (1971). *The Pivot of the Four Quarters*. Edinburgh: Edinburgh University Press.

Wilhelm, Richard and Cary F. Bayes, trans. (1967). *The I Ching or Book of Changes*. 3rd ed. Princeton, N.J.: Princeton University Press, Bollingen Series XIX.

Willis, Roy (1985). Do the Fipa have a word for it? In *The Anthropology of Evil*, ed. David Parkin (Oxford: Basil Blackwell), pp. 209-23.

Witzel, Michael (1984). Sur le chemin du ciel. *Bulletin d'Etudes Indiennes* 2, 213-79.

Wuilleumier, P. (1927). Cirque et astrologie. *Mélanges d'archéologie et d'histoire; Ecole Française de Rome* 44, 184-209.

Wyatt, N. (1986). Devas and Asuras in Early Indian Religious Thought. *Scottish Journal of Religious Studies* 7, 61-77.

Zaehner, R.C. (1955). *Zurvan: A Zoroastrian Dilemma.* Oxford: Clarendon Press.

Zhang Yachu and Liu Yu (1981-2). Some Observations about Milfoil Divination Based on Shang and Zhou *bagua* Numerical Symbols. *Early China* 7, 46-55.

Zimmermann, Francis (1987). *The Jungle and the Aroma of Meats: An Ecological Theme in Hindu Medicine.* Berkeley, Los Angeles, and London: University of California Press.

# Index